THE STORIES
OF YOUR LIFE

ABOUT THE AUTHOR

Ben Ambridge is a professor of psychology at the University of Manchester and the ESRC International Centre for Language and Communicative Development (LuCiD). His first two books, *Psy-Q* and *Are You Smarter Than A Chimpanzee?* have sold more than 100,000 copies across sixteen language editions. These books led to regular columns in *The Observer* and *The Big Issue*; appearances on Channel 4's *Sunday Brunch*, BBC Radio 2's *Chris Evans Breakfast Show* and the BBC's Tiny Happy People website; talks at Google, the Royal Institution, Oxford Literary Festival, Wilderness Festival and Bestival; and a TEDx talk, 'Nine Myths About Psychology, Debunked', which has had more than three million views.

THE STORIES OF YOUR LIFE

THE EIGHT MASTERPLOTS THAT
EXPLAIN HUMAN BEHAVIOUR

BEN AMBRIDGE

MACMILLAN

First published 2024 by Macmillan
an imprint of Pan Macmillan
The Smithson, 6 Briset Street, London EC1M 5NR
EU representative: Macmillan Publishers Ireland Ltd, 1st Floor,
The Liffey Trust Centre, 117–126 Sheriff Street Upper,
Dublin 1, D01 YC43
Associated companies throughout the world
www.panmacmillan.com

ISBN 978-1-0350-1838-3 HB
ISBN 978-1-0350-1839-0 TPB

Copyright © Ben Ambridge 2024

The right of Ben Ambridge to be identified as the
author of this work has been asserted by him in accordance
with the Copyright, Designs and Patents Act 1988.

All rights reserved. No part of this publication may be reproduced,
stored in a retrieval system, or transmitted, in any form, or by any means
(electronic, mechanical, photocopying, recording or otherwise)
without the prior written permission of the publisher.

Pan Macmillan does not have any control over, or any responsibility for,
any author or third-party websites referred to in or on this book.

1 3 5 7 9 8 6 4 2

A CIP catalogue record for this book is available from the British Library.

Typeset in Sabon LT Std by
Palimpsest Book Production Ltd, Falkirk, Stirlingshire
Printed and bound by CPI Group (UK) Ltd, Croydon, CR0 4YY

This book is sold subject to the condition that it shall not, by way of
trade or otherwise, be lent, hired out, or otherwise circulated without
the publisher's prior consent in any form of binding or cover other than
that in which it is published and without a similar condition including
this condition being imposed on the subsequent purchaser.

Visit www.panmacmillan.com to read more about all our books
and to buy them. You will also find features, author interviews and
news of any author events, and you can sign up for e-newsletters
so that you're always first to hear about our new releases.

TABLE OF CONTENTS

1. The Inside Story: Masterplot Recipes — 1
2. Quest — 11
3. Untangled — 35
4. Icarus — 57
5. Monster — 95
6. Feud — 122
7. Underdog — 156
8. Sacrifice — 185
9. Hole — 215
10. The Stories of Your Life — 245
    Notes — 257
    Acknowledgements — 277

1.

THE INSIDE STORY: MASTERPLOT RECIPES

This is a book about stories. So it seems only fair that I start by telling you mine.

A couple of years ago, I finally landed my dream job as a professor at the University of Manchester. But when I started out, I struggled. In the world of academia, two things matter: publishing journal articles, and getting grants (money to cover the costs of doing research). I was failing on both fronts: journals didn't want to publish my research, and funders didn't want to support it financially.

Worst of all, I had no idea why. The studies that I was running were sound. The peer reviewers – for both journals and funders – said so. The methods were described in detail. The statistical analyses were kosher. But my work was met with an overwhelming 'meh'.

After I'd been floundering for a while, a senior professor asked to take a look at my work. He put his finger on the problem almost immediately. 'You've got to find the story,' he said.

This was news to me. I had been working on the implicit assumption that only the content of my research mattered.

Ironically for a psychologist, I was ignoring decades of psychology research on how our brains store and process information: as narratives. As social animals, we humans have evolved to think not in terms of facts and figures, but stories and gossip; not science and statistics, but who's doing what to whom.

'Find the story,' my colleague told me. 'Pick a narrative, and stick to it all the way through.' I followed his advice, and it worked like magic. The research itself was the same, but once it was framed in terms of a story, the journal and funding-body reviewers could see what I was trying to do and why.

Which narrative did I use? That's the beauty of it: over the years, storytellers have served up a menu of overarching narrative arcs, or 'masterplots', to choose from. Having studied these masterplots, I was able to choose the one that best fitted the research project in each case. One time, it might be the Quest masterplot (Chapter 2), where the intrepid hero (me!) sets out to capture some hard-won, valuable knowledge that will cause us to re-evaluate everything we thought we knew. Another time, it might be the Monster masterplot (Chapter 5), where we must arm ourselves with a new weapon, device or technique to overcome some deadly threat (such as a hostile journal reviewer). Another time, it might be the Feud masterplot (Chapter 6), where well-matched, mirror-imaged rivals (here, competing theories) battle to the death.

It took me the best part of twenty years, though, to realize the fundamental importance of masterplots to human existence. At first, I saw them as little more than tools for writing academic papers, or curiosities to bring up when I found myself in conversation with film or literature buffs. In fact, they are so much more. What I eventually came to realize – and what I'd like to show you over the next couple of hundred pages – is the following:

THE INSIDE STORY: MASTERPLOT RECIPES

Whether we are aware of it or not, we humans attach the same set of predictable masterplots to just about all of the experiences that we go through in life, to the extent that we not only allow these masterplots to influence our decisions, but even – on occasion – manipulate them to achieve our goals.

So, what is a masterplot exactly? I hope you'll forgive me for reheating a metaphor that is rapidly approaching its use-by date, but the best way to understand masterplots – as they have long been used in all kinds of fiction – is to think of them as recipes for building stories. A recipe for, say, a sponge cake consists of a list of ingredients, and instructions for combining them in a particular order. Neither have to be followed to the letter. If you don't use exactly the right kind of sugar, or forget to sift the flour, your cake will still be edible. Certain ingredients (vanilla essence, lemon juice, a dusting of icing sugar) are optional, and an expert baker will often deviate from the recipe on purpose in order to achieve a particular result. But if you leave out one of the key ingredients, or put them together in a completely different order, you'll end up with an inedible mess.

It's the same for stories. The recipe for, say, an Underdog story (Chapter 7) consists of a list of ingredients (e.g. a put-upon hero from humble beginnings who eventually realizes their destiny) and instructions for combining them in a particular order (e.g. the humble beginnings at the start; the improbable, destiny-realizing triumph at the end).* Again, certain ingredients are optional (for example, the improbable triumph will often be a victory over a bitter rival, but it doesn't have to be), and an

---

* Literary critics sometimes call the ingredients the *fabula* and the order in which they're combined the *syuzhet*, but there's no need to memorize these terms, unless you want to impress your friends with your knowledge of 1920s literary theory; specifically, the branch known as Russian formalism.

expert storyteller will often deviate from the recipe on purpose in order to achieve a particular result. But, again, if you leave out one of the key ingredients (no improbable triumph), or put them together in a completely different order (the triumph at the beginning, rather than the end), you'll end up not with an Underdog story, but an inedible mess. This story recipe – consisting of a list of familiar ingredients and instructions for combining them in a particular order – is what I'm calling a masterplot.

How many masterplots are there, and what are they? My answer is eight: Quest, Untangled, Icarus, Monster, Feud, Underdog, Sacrifice and Hole. But this isn't the sort of question that has a single right answer. If we zoom in as close as we can go, there are as many plots as there are stories, since no two stories (or even two retellings of the same story) are identical. If we go to the other extreme, and zoom right out, looking down on all the world's stories from a great height, we can make out a single meta-masterplot that unites them all: the main protagonist is at home minding their own business, something happens that upends the normal order of things, then finally, for better or for worse, this unstable situation is resolved. This 'three-act structure'[1] – with its key ingredients of set-up, confrontation and resolution, always in that order – is the general recipe that all individual masterplots follow ('get the ingredients; mix them together; heat the mixture').

Over the years, various people have proposed different numbers of masterplots: seven (the journalist and author Christopher Booker); six (sci-fi author Kurt Vonnegut, mathematician Andrew Regan, computer scientist Marco Del Vecchio); ten (screenwriter Blake Snyder); or even twenty (film studies professor Ronald Tobias). These are all perfectly good ways of cutting the cake, but I like to think my eight hit the sweet spot, neither lumping together masterplots that are too different, nor splitting hairs between those that are all but identical. And while Booker, Vonnegut and the rest are concerned with identifying masterplots in works of fiction, I have a different – and much more ambitious – goal: the eight masterplots I set out here – Quest, Untangled, Icarus, Monster, Feud, Underdog, Sacrifice and Hole – cover not just fiction, but real life, too.

Understand these eight masterplots, and you'll go a long way to understanding many aspects of human behaviour, including your own. Why did he say that? What is she trying to prove here? Why is he so determined to treat her as an enemy? Why did they move *there*, and why *then*? Why did I treat them so badly – am *I* the baddie? Biologists have a saying: 'Nothing in Biology makes sense except in the light of evolution.' Well, this psychologist has a saying too: 'Nothing in human psychology makes sense except in the light of masterplots.'

This is a bold claim, so before we get to the masterplots themselves, let me explain. Key to understanding the importance of masterplots is understanding where these recipes came from in the first place. Like culinary recipes, they weren't handed down by God, or created by some lone genius. Instead, they emerged gradually in the collective human consciousness as storytellers, or 'content creators' as we might call them today, intuitively – and probably by luck as much as judgement – figured out what their listeners, or 'consumers', found both satisfying and delicious. The reason that these particular eight masterplots hit the sweet spot is that they capture, and distil to their essence, human experiences that were *already familiar* to the earliest listeners. Just like modern-day listeners – and perhaps even more so – these early listeners would have brought with them lived experience of feud, sacrifice, underdogs and all the rest. Thus, a story that has been crafted according to one of the eight masterplot recipes resonates deeply with us still, because it is – in a deep and personal sense – a story that we already know.

But why does this matter? Doesn't familiarity breed contempt? Why do we prefer personally familiar, predictable stories to new and fantastical ones? There are two reasons, but both boil down – ultimately – to Darwinian evolution by natural selection (sometimes called 'survival of the fittest'). First, stories – in the form of gossip – allow us to make (usually) accurate predictions about the behaviour of others; for example, that a particular individual doesn't keep their promises. Individuals who memorize and internalize these stories, and use them to make correct predictions about the behaviour of others, are at a clear advantage when it comes to natural selection.

Second, recent discoveries in neuroscience[2] have provided powerful support for an idea that can be traced back to the work of the German physicist Hermann von Helmholtz, writing in the 1860s: the brain is essentially a prediction machine.[3] Whatever you're doing – strolling around the park, having a conversation, listening to a story or a piece of music[4] – your brain is constantly trying to predict, millisecond by millisecond, exactly what is going to happen next (e.g. 'That dog looks friendly – I think he's coming over to say hello'). When it gets it right, the brain rewards itself with a hit of pleasure,[5] similar to that triggered by sex, food, winning a bet or just about any pleasurable activity.* Again, it's all about natural selection: a brain that is constantly making incorrect predictions ('That lion

---

* Many books will tell you that the neurotransmitter dopamine is 'the pleasure chemical', and it's this that you're getting a 'hit' of in these and other pleasurable scenarios. The reality is much more complex – and still debated – but essentially dopamine is more to do with motivation or wanting; 'pleasure' per se is more to do with endogenous opioids (opioids that are produced by the body as opposed to artificial opioids like heroin).

looks friendly'), or that sometimes fails to make any prediction at all, isn't going to be around for long.

Masterplots, then, give the brain a helping hand. As soon as we recognize the type of plot that is unfolding, we can start to make predictions about what will happen next and how the story will ultimately end. We just can't help ourselves from actively predicting the outcome, chasing the chemical hit that we get when we're right. That's why a 2011 study at the University of California, San Diego found that 'Story Spoilers Don't Spoil Stories'.[6] That's right – the researchers found that for mysteries, ironic-twist stories and 'literary' stories alike, readers actually gave *higher* enjoyment ratings when the stories had been 'spoiled' with a summary beforehand. Indeed, despite the modern preoccupation with 'spoiler alerts', it was well over a hundred years ago that Anton Chekhov offered the famous advice, 'If in the first act you have hung a pistol on the wall, then in the following one it should be fired. Otherwise, don't put it there.'

Our brains, then, love predictability. But there's a twist: when something is *too* predictable, the pleasure hit is reduced.[7] This, too, makes sense in evolutionary terms, because a brain whose predictions are correct 100 per cent of the time can't learn anything. Just like a student in a classroom, your brain learns when it gets something wrong ('7 x 6 = 40') and is corrected by a teacher ('No, it's 42'), in your brain's case, the teacher being unfolding events in the world. What we want in a story, then, is this sweet spot of 'surprising familiarity', of 'unpredictable predictability'. We know what's going to happen: the hero of our Quest (Chapter 2) is going to make it home. But we don't know *how* it's going to happen; the journey will be full of surprising, unpredictable events. If events are too predictable (the hero hops on the next direct flight home) or too unpredictable (they turn into a frog and leap off the face of the earth), they don't make a satisfying story, in just the same way that no (edible!) recipe would consist solely of salt, flour and water, or a totally random selection of ingredients (ham, milk, peanut butter, flour and orange juice).

Masterplots, then, allow storytellers to create stories that are – in an important sense – already familiar to their listeners, and that have the perfect balance of neurochemically rewarding unpredictable predictability. But masterplot recipes aren't just for fiction writers. All of us view real-life events through the lens of one or other of these eight masterplots, allowing them to manipulate us, and using them to manipulate others.

Why? The answer is that fictional and real-world plotlines are constantly reinforcing one another. We have already seen that these eight masterplots emerged in the first place because they retell stories that are already familiar to us from our real-world experience. Young children are – not to put too fine a point on it – terrible at both following narratives and constructing their own, but, as anyone who has read *The Gruffalo* hundreds of times to a four-year-old will testify, they love stories, and they love *repeated* stories.[8] By the time we reach adulthood, we will each have internalized literally thousands of stories, between them exemplifying each of the eight masterplots. Precisely because these masterplots strike us as familiar and – surface details aside – true to life, we cannot help but interpret real-world events through their lenses.

None of this is news, of course, to propagandists who have long understood the importance of what we would today call 'controlling the narrative'. Once you realize that masterplots shape *your own* behaviour, it is a short step to realizing that you can control *other people's* behaviour by deliberately framing events according to whatever masterplot suits your goals. Beware the politician who casts himself as the hero of a Monster story, when it's actually him that's the monster; or the brand that tries to sell you an Underdog story, when their product sprang fully formed from a committee of rich execs. A particularly current and dramatic example, which we will explore in detail in the final masterplot chapter, is the ongoing battle over the narrative framing of climate change, a battle which – it is absolutely no exaggeration to say – will determine the future existence of our species.

But we are getting ahead of ourselves. Before we set sail with our first masterplot (Quest), I'd like to orient you with a brief road map. For each masterplot, we'll start off by looking at a work of fiction that follows the relevant recipe more-or-less to the letter; it could be a novel, a play, a movie, a TV series or even a poem. We'll then be in a position to set out **The Masterplot Recipe**, in terms of both the key ingredients, and when and where each must be added to the story. Next (in **Stranger than Fiction**), we'll see the recipe in action in real life with a true story; it might be that of a musician, a politician, a CEO, a brand, a sports team or just an ordinary granny. With both fiction and non-fiction examples under our belt, we'll delve into **The Science Behind the Story**, and investigate the psychological underpinnings of the relevant masterplot. Next and most crucially, in **Under the Influence** sections we'll explore the effects of the masterplot on human behaviour, real-world cases where the relevant masterplot recipe doesn't just *describe* human behaviour, but actively *shapes* it. This will lead us to consider cases in which the relevant masterplot has been abused, misused or perverted for nefarious ends: **Plot Twisted**. But it's not all bad news; we'll end our discussion of each masterplot on an upbeat note: **Happy Endings**. Yes, masterplots are tools for manipulating human behaviour, and like all tools they can be put to evil uses;

but if we think bigger and brighter and better, they can be catalysts for human progress.

Because they distil the very essence of the human condition into super-concentrated form, masterplots can do anything. They can dupe, trick or mislead, but they can just as easily inspire. As I will show, they have already helped put a man on the moon (Quest), fight starvation (Sacrifice) and defeat addiction (Monster); they've ended blood feuds (Feud), given comfort to grieving parents (Icarus) and divorcing couples (Untangled), and helped a sports team triumph at odds of 5,000:1 (Underdog). But they can do so much more. If we can master their unique power, masterplots can help us push the boundaries of scientific discovery, fight for a just world, save lives and – just maybe (Hole) – prevent the impending extinction of the human race.

Right now, of course, you haven't embarked on any of these adventures; you're just sitting at home, reading a book, minding your own business, when you receive an intriguing invitation. Grab your keys and your phone, and put on your coat – we're headed off on a Quest . . .

2.
QUEST

What's the greatest story ever told?

When the BBC polled over a hundred literary critics, one story came out head and shoulders above the rest: *The Odyssey*.[1] Even if, like most people, you haven't read Homer's epic, I'm sure you're familiar with the general idea of an odyssey as an epic journey; from the Nintendo Switch game *Super Mario Odyssey*, perhaps, or the classic movie *2001: A Space Odyssey*.

What's so great about *The Odyssey*? It can't be the writing. The original poem is written in Ancient Greek, and comes to modern readers in the form of translations that are criticized as either old-fashioned and stuffy or – as with Emily Wilson's recent version – not old-fashioned and stuffy enough. No, what has kept this story at the top of the critics' charts for almost three thousand years is just that: the *story* . . .

We join *The Odyssey* close to the end, with most of it told in flashback (a trick still used today, for example in Quentin Tarantino's *Once Upon a Time in Hollywood* or Martin Scorsese's *The Irishman*). Our hero, Odysseus, has finished fighting in the Trojan War, as documented in another Homerian epic, *The Iliad* (in which he plays a minor role, making *The Odyssey* something of a spin-off sequel; *The Il-iad*, if you will). But, although ten years have passed, he still hasn't made it home to Ithaca, where his son, Telemachus, and his loyal wife, Penelope, are waiting, the latter constantly fighting off suitors. As the story opens, Odysseus is living on the isolated island of Ogygia, as a prisoner of the goddess Calypso. She offers him immortality if he'll marry her, but he refuses – although he sleeps with her, obviously (as Emily Wilson's translation puts it, he's what we would today call 'a complicated man'). Eventually, the gods intervene: Athena pleads Odysseus's case to the king of the gods, Zeus, who sends his messenger, Hermes, to tell Calypso to release our hero. She does, and he sails off, getting as far as Phaeacia, whose citizens make up the audience for the flashback retellings of his exploits . . .

First, he and his crew went to the island of the lotus-eaters, where this mysterious fruit proved so delicious that some crew members fell into a trance and had to be dragged back to the ship. Next, they wandered into the cave of a Cyclops – a one-eyed giant who started eating them. But Odysseus fooled him by giving his name as 'Nobody', then got the Cyclops drunk and attacked him with a wooden stake. When the neighbours turned up to see what all the fuss was about, the Cyclops shouted, 'Nobody is attacking me', and they went back to bed. Odysseus's next run-in was with Circe, a witch-goddess who turned his men into pigs – although he still slept with her, obviously. For a *year*. Recognizing his weakness, Odysseus had his men tie him to the ship's mast as they sailed past the island of the Sirens, famous for luring sailors to their death with their spellbinding song. Edging between a giant whirlpool (Charybdis) and a six-headed monster (Scylla), the crew steered too close to the latter and lost one member for each head. Having reached land at Thrinacia, the hungry sailors hunted cattle belonging to Helios, the god of the sun, and were all drowned as a result. Only Odysseus survived, and washed up on Ogygia with Calypso.

QUEST

Impressed by these exploits, the Phaeacians take Odysseus back to Ithaca where, disguised as a beggar, he competes for Penelope's hand in marriage in an archery contest. He wins with a shot that, famously, only Odysseus himself could have pulled off, firing his arrow through a row of twelve axes. Having thus revealed his true identity, he kills the suitors and is finally reunited with his wife and son. Phew!

THE MASTERPLOT RECIPE

As one of the oldest stories that is still read today, *The Odyssey* doesn't just *follow* the Quest recipe:[2] it played a major part – along with *The Iliad* – in establishing this recipe in the first place, which has been followed by just about every quest story since, from *The Pilgrim's Progress* to *The Super Mario Bros. Movie*; from *The Divine Comedy* to *Watership Down*; from *The Lord of the Rings* to *Barbie*.

Although stories that follow the Quest recipe have a distinctive start and ending – which we'll get to in a minute – most of the flavour comes from the ingredients that are added (in any order) during the main, central part of the story: the quest or voyage itself. A particularly key ingredient is our hero's encounter with **monsters** (occasionally metaphorical; but most often literal). In *The Odyssey*, it's the Cyclops; in *The Super Mario Bros. Movie*, it's Bowser; in *The Lord of the Rings*, it's Sauron, the Watcher, the Balrogs, the Fellbeasts (there are loads, OK?). More often than not, our hero will have to overcome some **great temptation**. The ring, as in *Lord of the*, is of course temptation personified (ring-ified?). For Barbie, the temptation is to get rid of her cellulite by – clunking metaphor alert – getting back into the box offered to her by Will Ferrell and the rest of the Mattel executives. For our 'complicated' hero Odysseus, the temptation is any female with a pulse (they don't even have to be human). At some point, our hero is usually trapped **between a rock and a hard place**; forced to choose somehow between two options that are equally bad. Odysseus tries, and fails, to navigate his way in between the giant whirlpool and the six-headed monster. In *The Pilgrim's Progress*, the titular Pilgrim – the imaginatively-named Christian – has to navigate his way between a ditch and a bog. **The supernatural** almost always makes an appearance at some point. Mario has his Super Star; Barbie has Weird Barbie; to say nothing of the quests – *The Pilgrim's Progress*, *The Lord of the Rings*, *The Odyssey* – in which the examples are too numerous to mention. Helpers are also a particularly key ingredient in this recipe. In fact, I challenge you to come up with a single quest story in which our hero is unaided. Odysseus has both **travelling companions** that come along for the ride (like Frodo's Samwise or Mario's Luigi) and **local helpers** such as the Phaeacians (like Frodo's Strider or Mario's toads).

The final middle-of-the-story key ingredient – and the one that really gives the finished product its texture – is **unworldliness**. That is, the place that our hero goes to is *weird*. It might be a literal other world – the Mushroom Kingdom (*Mario*), the 'real world' (*Barbie*), Middle Earth (*The Lord of the Rings*), Wonderland (*Alice's Adventures in . . .*), Oz (*The Wizard of . . .*); or it might be a remote island – Ogygia (*The Odyssey*), Lilliput (*Gulliver's Travels*), The Island of Despair (*Robinson Crusoe*). The point is, it's somewhere very different from home; somewhere where the rules – often, even the basic laws of physics – don't apply. This is the part of the movie that they put in the trailer and on the poster. Just what would it be like to be . . . *there*?

While the ingredients we've tasted so far define the central part of the Quest masterplot – the quest itself – the recipe has three further ingredients that must each be added in exactly the right place. First of all (flashbacks notwithstanding) our hero must be summoned by a **call to action**. Our hero is just sitting around minding their own business when – wham – Zeus intervenes, Barbie's feet go flat (you'll have to watch the movie to get that one), or Mario and Luigi get sucked into a warp pipe (ditto). There is no indecision, no shilly-shallying around; the hero just *has to* go. Now.

Second, all of the action with the monsters, the temptations, the supernatural and so on must culminate in a **final ordeal**: Odysseus's archery contest, Mario's battle with Bowser, or Frodo's with Sauron (or, some would argue, the ring itself). Third, the story ends not with – despite its name – the final ordeal, but with the **life-renewing goal**. Homer could have had Odysseus finish his journey in some fabulous unearthly paradise. But he doesn't. Instead, he ends up back home with his wife and son. It's the same in *Alice's Adventures in Wonderland*, *The Wizard of Oz*, *Robinson Crusoe*, *Watership Down* . . . Even the exceptions prove the rule: having returned home to Brooklyn in triumph, Mario and Luigi move to the magic kingdom, but take jobs as plumbers. Although Frodo later heads off for the Undying Lands, he and the other Hobbits initially go back to the Shire, where Sam gets married. Barbie settles in the real world, but – at least in one respect ('I'm here to see my gynaecologist') – is unable to escape her plastic origins.

The point is, none of these fictional heroes simply ends up back where they started. Yes, Odysseus has come full circle. But he had to go away in order to come back, bringing with him something that would allow everything to start anew; something that would demonstrate that what he was really looking for had been there all along: the love of his wife and son. It's not 'as you were' but 'the new normal'; not simply repairing, but building back better.

STRANGER THAN FICTION

Karl Bushby – I don't think he'll mind me saying – is not a literary man. Having struggled with dyslexia, he left school at sixteen to join the army. Eleven years later, he left.

Drifting in the army, Karl got a birthday card from his father, a former army man himself, which mentioned in passing that a couple of Special Forces guys were planning to walk from London to New York over the Bering Strait. The Bering Strait is a narrow body of water – just over fifty miles wide at its narrowest point – which separates Siberia from Alaska. In winter, it freezes, at least partially, meaning that it is theoretically possible to walk, hop and swim from one side to the other. This was Karl's irresistible **call to action**, or, as he put it to me, 'the missing piece of the puzzle, connecting the Americas to Asia'. Karl had long had a vague ambition to walk around the world. Now this vague ambition was a theoretical possibility.

In November 1998, Karl set off on foot from Punta Arenas, on the southernmost tip of mainland Chile, for home: Hull, a city in the north of England. Yes, like Odysseus, and the hero of just about every quest story, Karl's goal was – and, at the time of writing, still is – to return home.

Like many children of his generation, Karl grew up with Tolkien's Middle Earth books, of which *The Hobbit* was a particular favourite. These books sparked what Karl calls a 'yearning for journeys that had always been there', and that, in fact, is probably there in most of us. Does any child, when hearing the Tolkien stories, *not* wish that they were heading off on such a quest themselves? The difference is that Karl actually acted on it.

While a fictional voyage-and-return story might have inspired Karl's quest, for over two decades now he has been living one. What is so striking about Karl's story – almost eerie, in fact – is the way that it incorporates just about every single ingredient in the masterplot recipe. The **local helpers** are the – by Karl's estimation – 99.9 per cent of people he met along the way who didn't know him, but were happy to offer him a meal, money or a place to sleep.

Travelling companions played little role in the first part of the trip – the trek through South, then Central, then North America – but became crucial when Karl finally reached the Bering Strait and teamed up with seasoned explorer Dimitri Kieffer. Together, in March 2006, the two successfully made the east–west crossing, and to date remain the only people in the world to have done so. **Temptations?** Sure, by my reckoning, Karl (helped along by his uncanny resemblance to Kurt Cobain) has had more romantic encounters than even Odysseus. And, just as for Odysseus, although these liaisons often held Karl up for a few weeks or months, they never came close to scuppering his return home. **The supernatural** made an appearance (sort of) in the deserts of Chile, where Karl saw a rotating, oscillating bright light hanging in the air, subtly changing shape. A UFO? Alas, no, just a white plastic bag caught in a thermal current between two hills. That said, like Odysseus, Karl had an underworld experience in Colombia, where he slept in a graveyard alongside open graves.

As far as **monsters** are concerned, Karl encountered plenty in the Darién Gap: a 100-km-long stretch of jungle that straddles Colombia and Panama, infested with wild animals, poachers, drug smugglers and the notorious FARC, the Revolutionary

Armed Forces of Colombia: Communist guerrillas, perpetually at war with the Colombian Government, and big players in the drug business (who you might be familiar with from the Netflix series *Narcos*). Karl evaded the FARC by staying off the roads, floating down crocodile-infested rivers and slashing his way through thick jungle (clearly a large serving of **unworldliness** for a man who grew up in Hull). As if this wasn't enough, Karl also had to dodge – at various points along the way – both snakes and polar bears. Yet he maintains that the real monsters in his story are governments and their agencies, particularly the Russian Federal Security Service, the FSB. When Karl and Dimitri became the first people to cross the Bering Strait into Russia, their reward was detention and, later, a five-year ban from the country altogether. The pair were released only when a deal was brokered between – you couldn't make this up – John Prescott (the then Deputy Prime Minister of the UK, and MP for Karl's home town of Hull) and former Chelsea Football Club owner Roman Abramovich (the then-governor of the Chukotka region). And even then, Karl's ban on entering Russia remained in place until he flew back to the west coast of the United States and walked the breadth of the country to plead his case at the Russian Embassy in Washington DC.

When I first Zoomed with Karl back during Covid lockdown (now that was an otherworldly experience, right at home) he was trapped between his **rock and a hard place** (for Odysseus, Charybdis and Scylla). After crossing the Bering Strait and travelling west through Russia, Mongolia, China, Kyrgyzstan, Uzbekistan and most of Turkmenistan, Karl was unable to enter Iran, due to a combination of Covid restrictions and Western sanctions. If Iran was Karl's Charybdis, then his Scylla was the alternative route via Uzbekistan, Kazakhstan and Russia, a country that has not only caused Karl no end of visa trouble, but has seemingly also leaned on its neighbour Kyrgyzstan to do likewise. In any case, any lingering possibility of travelling through Russia was well and truly quashed when Putin launched his invasion of Ukraine in 2022.

Undeterred (well, quite deterred, but definitely not giving up), Karl avoided both Iran and Russia by heading back into Uzbekistan (forgetting about Turkmenistan altogether) and up into Kazakhstan. The plan, at the time of writing, was for Karl to swim 259 km across the Caspian Sea to Azerbaijan;[3] as Karl told me, 'Not really something I signed on for at the beginning of this endeavour, but hey, whatever it takes, right?' If this part comes off, though, Karl will be only around 5,000 km from home, with almost all of it on the home terrain of Europe. A **final ordeal**? We'll have to see. Karl's plan is to return to the UK by walking through the Channel Tunnel from France. Will Brexit Britain pull the necessary strings, or pull up the drawbridge? I'm sure Karl will swim the Channel if he has to.

But, as we've already seen, the final ingredient in the Quest recipe isn't the final ordeal – it's a denouement in which the hero returns to his old life, but on surer footing, having brought back something that allows him to begin anew. Karl's **life-renewing goal** is to work full-time on a non-profit organization that he's currently setting up to improve scientific literacy. What is Karl bringing back with him that will kick-start this new chapter? Two things: first, an appreciation of the size of the task facing him: 'Travelling the world, you see it [scientific illiteracy] in every nook and cranny'. He calls the coronavirus pandemic a 'classic case: It's hard to work together and fight something on this scale when 40 per cent of a nation, let's say for example the US, is not even willing to accept that it even exists'. Second, 'a rediscovered faith in humanity': 'Having received so much kindness on the road . . . this trip has reaffirmed my love affair with humanity – we're an incredible species and we just have so much potential. Humanity is just the most incredible thing if we just get it right.'

THE SCIENCE BEHIND THE STORY

Karl's story raises an important question: why do we have to go away and come back to find the metaphorical holy grail that was eluding us? Why did Karl discover a passion for science communication only after he'd walked three quarters of the way around the world? More generally, one thing that's distinctive about the Quest masterplot is the importance of place: the need to physically *go somewhere*. But *why*?

In the summer of 1997, having just completed my first year of university studying French and Russian, I set off on a month-long trip to Yaroslavl and St Petersburg, organized by the university to give our rudimentary Russian the kick-start it needed. I had the time of my life: drinking vodka in the park with Russian teenagers, being booted off the local five-a-side pitch by garishly tracksuited mafiosi, dancing in the most unfashionable nightclubs in the entire world (guarded by those same mafiosi), watching the local football team (Shinnik – the tyre-makers), eating mystery meat with writing on it . . . But as soon as I came back to England, I switched my university course from French and Russian to Psychology. Fantastic though my trip was, it made me realize that what I loved wasn't so much Russian and Russia (or French and France), but the inner workings of our minds that made these languages possible (in fact, my high-school French teacher had always said that I should study linguistics). Having restarted university as a psychology student, I was quickly drawn to the area of psychology that looks at language – how we learn it as children and represent it as adults – where I have remained to this day. And that's just *my* story. I'm sure if you think about it for a minute or two, you can think of a similar quest or 'voyage of discovery' that changed the course of your life story in some way.

Now, could I have come to this realization without my trip to Russia? Logically, yes, I *could* have: I had all the same information available to make the decision. But I *wouldn't* have. So why do we need to go away and come back to make such decisions? The answer is that the particular location in which you find yourself isn't just a backdrop to the actions taking place onstage; it's a crucial part of the story itself. This was shown clearly in a famous 1975 study conducted at the University of Stirling in Scotland. Hardy members of the Clydebank, East Kilbride and Stirling diving clubs donned their gear and plunged into the chilly crystal-clear waters of the River Clyde. Once fully submerged, they either attempted to recall a set of words that they'd been taught on dry land or – vice versa – attempted to memorize a set of words for later recall once out of the water. The experiment did not always proceed smoothly. 'One diver,' notes the paper dryly, 'was nearly run over during an underwater experimental session by an ex-army amphibious DUKW'. Nevertheless, the results were clear: words that had been learned on land were recalled better on land than underwater, and vice versa. In fact, switching the location reduced the number of words recalled by around a third.*

---

* If you'd like to try for yourself – perhaps on the sofa versus in the bath – here's a list of words similar to that used in the original study: admit, angle, banner, cherry, despise, error, farmer, funeral, heavy, ideal, jewel, journey, kitten, knowing, lady, language, madam, normal, outline, over, police, pocket, portion, question, quiet, railway, record, separate, trifle, ugly, voyage, welfare, witness, writing, yellow, youthful.

Why? The key is learned associations. Just as Pavlov's dogs famously learned to associate a bell with the delivery of food, so these divers learned to associate the underwater environment with one set of words and the riverbank with another. As we saw in Chapter 1, the brain is a prediction machine. When presented with, for example, the word *admit* underwater, divers unconsciously learned that the surrounding cues – the water, the cold, the mud – predicted the appearance of that word, just like Pavlov's bell predicted the arrival of food. When taking the recall test underwater, the divers used the water, cold and mud to predict – that is, to bring to mind – the word *admit*, just like Pavlov's dogs used the bell to bring to mind the food. But when taking the recall test on dry land, they couldn't use any of those cues.

In just the same way, when you're on home turf, you can't shake off well-learned associations. Your alarm goes off, so you pick up your phone to silence it, and check (depending on your age) TikTok, Instagram, Facebook, X (you know, Twitter), or whatever eventually kills it off (Bluesky? Mastodon? Surely not *Threads*?!). You feel the cold of the kitchen floor on your bare feet, which is your cue to fill the kettle. The sound of the running water is your cue to think about what you have planned for your day . . . Changing your routine isn't impossible, of course, but it's difficult and requires conscious effort. Now think about what happens when you go away. Say you're on holiday, so you didn't set an alarm. Maybe you don't check your socials, but instead turn on the news. Instead of filling the kettle, you look on your phone for a local cafe. Once there, you get chatting to someone who suggests somewhere to go that day . . . suddenly, you've changed your whole routine without even trying.

And here's the bonus: if you're away for long enough, the new associations you've learned will start to chip away at the old. Can you recreate in your mind's eye that feeling you get when you return from a long trip and look around your house with a kind of wide-eyed astonishment? Everything is entirely familiar, of course, but at the same time, somehow impossibly new. That's what prediction error – those incorrect predictions that lead to learning – feels like. Now when you open the front door, your brain predicts the hallway of your holiday let, but what's this? A completely surprising – though entirely familiar – hallway. Maybe waking up now predicts something other than checking your socials. Maybe the cold of the kitchen floor now predicts something other than running the kettle. You're back where you were, but things have changed. And so have you. You had to go away to come back.

You might be thinking this sounds a bit far-fetched. Yes, of course it's easy to slip into familiar habits, but can't we change them with just a bit of mind over matter? Surely we don't have to go to all the bother of going somewhere else entirely just for things to be a bit different when we come back?

In 1982, a Canadian psychologist named Shepard Siegel published the findings of a ground-breaking experiment conducted with rats.[4] Every other day for fifteen days, each rat was injected with a dose of heroin (which, by all accounts, they thoroughly enjoyed). Finally, each rat was given a particularly large dose of heroin – almost double the biggest they had had so far. The twist is that half of the rats received this large dose in the same cage where they always got their fix; half received it in a different cage in another room. The rats who had this bigger dose in their familiar 'drug-taking' environment tolerated it relatively well, with more than two-thirds surviving. But the rats who had this bigger dose in a different environment weren't so lucky, with around two-thirds of them dying from the overdose. In the familiar drug-taking cage, the brain uses the cues of that particular environment to predict – there's that word again – that some heroin is coming, and begins to make the necessary adjustments. In the new cage, the brain is unable to predict the impending heroin, and doesn't know what's hit it until it's too late. And it's not just rats. Siegel's study was inspired by reports of human heroin users overdosing after taking a quantity of the drug that they were well-accustomed to, but in surroundings to which they were not.

If learned associations, learned predictions, are strong enough to protect you against an otherwise fatal dose of heroin, we should not be surprised that they are strong enough to 'protect' you against your best efforts to change your routine. It's only by changing the predictors – changing your surroundings – that your brain can start to learn new associations. With drug addictions, smoking and eating disorders, of course, this isn't easy to do, because drugs, cigarettes and doughnuts have lots of different 'cues' or 'predictors', many of which can be difficult or impossible to avoid: the time of day, your own mental state, your family. But the more of these you can avoid, the better your chances of breaking the link; of breaking the prediction; of changing the narrative.

UNDER THE INFLUENCE

Perhaps more than any other, the Quest masterplot causes people to do things that would never have even occurred to them otherwise, purely out of a desire to be the hero of a real-world Quest story.

Walking the entire way around the world is painful. It's dangerous. It's expensive. Above all, it's *mind-numbingly tedious*. It might sound exciting and glamorous to us outsiders but, as

QUEST

Karl told me, 'When you're the guy actually doing this, it gets incredibly monotonous. It's an office job: another 30k down a road through Kazakhstan isn't the most exciting lifestyle at all . . .' And for what? A flight would get Karl home in a fraction of the time for a fraction of the cost. There is literally no point.

Except, of course, there is. It's not just that Karl's journey happens to *follow* the masterplot recipe. It's not even that Karl *uses* the Quest masterplot to motivate himself. It's much more: Karl's walk *is* a Quest narrative; without that narrative, it's nothing.

We see the same thing on a smaller scale all the time. Almost every local news bulletin features an item about a modern-day Odysseus who is climbing a mountain or traversing the length of the country for charity. Take, for example, Paul Taylor, who decided to tour the UK's rudest place names. Starting at Shitterton in Yorkshire, he sampled the delights of Twatt in Orkney, Butthole Lane in Leicestershire and the New Forest's Sandy Balls Holiday Park; or – posthumous controversies aside[5] – Captain Tom Moore's quest to walk 100 lengths of his garden during the UK Covid lockdown. For as long as there has been religion, there have been pilgrimages – literal quests that include most of the ingredients of the masterplot: the irresistible call to action, the supernatural, the companionship, the kindness of strangers and local helpers, the triumphant arrival, and finally the return home, changed for the better.

In fact, I would go so far as to suggest that our internalization of the Quest narrative largely determines why some feats strike us as admirable, while other equally challenging feats strike us as pointless. On the small-scale, everyday level, few of us would complain about donating to a friend or relative who is doing a sponsored walk, particularly if it takes in an exotic or far-flung location or, in one popular variant, involves travelling as far as possible with no money. But nobody would expect sponsorship for completing tasks that are just as difficult – and arguably worthier – such as learning Chinese characters or mastering algebra. The lack of identifiable Quest ingredients leaves us cold.

The same is true for feats on a grand scale. When asked why he wanted to climb Everest, the British mountaineer George Mallory is famously supposed to have replied, 'Because it's there.' But an infinity of other challenges are 'there' too. Why this one? 'We choose to do . . . [these] things not because they are easy, but because they are hard,' said President John F. Kennedy in 1962, vowing to put a man on the moon. Well, yes, but ending racial discrimination or child poverty or world hunger are also hard things – why not those? What Kennedy grasped was that the more ingredients of the Quest narrative you can throw into the mix, the more the endeavour will capture the public imagination.

If you think ending racial discrimination, child poverty or world hunger are pie in the sky – not just hard, but impossible – contrast the moon landings with some things that we have achieved since: the first 'test tube baby' (Louise Brown in 1978), the eradication of smallpox (1979), the world wide web (1993), cloning a sheep (Dolly, in 1996), sequencing the human genome (completed in 2003), coming up with a chat-bot that shows human-like command of language (ChatGPT in 2022). These achievements have far bigger implications for our day-to-day existence than the moon landings, and the technical challenges involved were, for the most part, greater. For example, according to one estimate,[6] training ChatGPT required 20,000 processing units each with 80 gigabytes of memory; 400,000,000,000 times more than the guidance computer used in the Apollo 11 mission (4 kilobytes). Yet none of these subsequent achievements captured the public's imagination in the same way as the moon landings. And this is just the achievements that made the front pages. Other achievements that are equally impressive from a technological perspective[7] – finding the Higgs boson (2012), detecting gravitational waves from black holes (2015), a vaccine for Ebola (2016) – barely got off the science pages.

The same is true for hypothetical future achievements; what do Elon Musk and the other tech bros say when they want attention? 'We're going to invest heavily in a range of technologies aimed at slowing climate change'? *Boring.* 'We're going to colonize Mars'? *Now you're talking.* What the other achievements lack, but the moon landings (and hypothetical Mars colonies) have in abundance, are the key ingredients of the Quest masterplot, most obviously the ingredient of **otherworldliness**; travelling to a place that is nothing like Earth, where the usual rules – even gravity – don't apply.

A few years ago, I had the opportunity to meet Peter Moore, then CEO of Liverpool Football Club. Although I hadn't properly started work on this book, masterplots were very much on my mind, so I explained the general idea, and asked him if he'd ever found himself being influenced by one. His answer crystallized the power of masterplots to such an extent that I knew straight away I would have to write this book. Moore, originally from Liverpool, had moved to America when he was only just out of his teens, trying to make it as a professional footballer (the English kind) or – failing that – a coach. Eventually, he found his way into video games, leading the launch of the Sega Dreamcast and the Microsoft Xbox, before becoming Chief Operating Officer at Electronic Arts (makers of the FIFA games). In 2017, Liverpool FC came calling. It was a fantastic opportunity. But Moore was settled in California. His kids had been born there. They liked it. He liked it. How could he persuade them, and himself, that this was the right thing to do? He did it, Moore told me, by taking a conscious decision to reframe his move to Liverpool as a return home. He didn't feel that way at first: sunny California, not rainy Liverpool, was home. But he knew he could succeed only by selling the move, to himself and his family, as the return leg of his Quest. He did, and oversaw the club's most successful period since the 1980s, winning the Premier League, the Champions League and the FIFA Club World Cup. Such is the power of the Quest masterplot.

PLOT TWISTED

No wonder, then, that it's a masterplot that is ripe for perversion. The Quest gone wrong is a fictional trope almost as old as the Quest itself. First published in the 1600s, Miguel de Cervantes's *Don Quixote* (which gives us both the phrase 'tilting at windmills' and the adjective 'quixotic') is widely regarded as not just the first example of this genre, but the first modern novel of any kind.

Don Quixote de la Mancha, who starts out as plain old Alonso Quixano, loses his mind after reading too many books on ancient chivalry and decides (**call to action**) that he is a knight himself. He starts off at the local inn, which he decides is a castle (**unworldliness**), taking a local farm girl and a landlord to be – respectively – Lady Quixote and the lord of the castle. After recruiting his neighbour as his squire (**travelling companions**), he attacks the famous windmills, which he takes to be giants (**monsters**). He also attacks a pair of monks (**between a rock and a hard place**) whom he takes to be wizards holding a woman captive. Having picked up some nasty injuries in a fight at another castle (inn), our Don tends his wounds with a treatment of his own concoction, which of course only makes things worse. Next, he is tricked by a woman pretending to be a Guinean princess and (**final ordeal**) fights the giant who stole her kingdom (actually, some wineskins in yet another inn). The first book ends with Quixote – believing himself to be under a spell (**the supernatural**) – being dragged home by his friends. A sequel, in which he fights a fake Don Quixote who himself is from an unauthorized sequel (perhaps the first ever example of fan fiction) is – believe me – more of the same.

If he were alive today, rather than a fictional character from the 1600s, Don Quixote would presumably qualify for a diagnosis of what the American Psychiatric Association's *Diagnostic and Statistical Manual of Mental Disorders* calls 'Schizophrenia Spectrum and Other Psychotic Disorders', two major symptoms of which are delusions and hallucinations.[8] Although *Don Quixote* may sound far-fetched, so powerful is the Quest masterplot that, given the right circumstances – or, perhaps more accurately, the wrong circumstances – the brain of a person experiencing psychosis will sometimes attach this narrative to a series of meaningless and quite unconnected effects. Tom Hartley, now a psychologist at the University of York, published a fascinating article describing an episode of psychosis that he had experienced thirty years earlier, when he was still a young student (though, fortunately, as for around half of those who experience such an episode, it was a one-off event, rather than the prelude to a condition such as schizophrenia).[9]

Interestingly, when I explained what I was up to with this book, and asked Hartley to pick out his favourite masterplot recipe from my menu of eight, it was Quest – alright, along with Underdog – that he chose. Is it too fanciful to suggest that, when presented with a series of unrelated events but somehow compelled to stitch them together into a narrative, we fall back on our favourites; those that feel the most comforting, the most familiar or just the most entertaining?

Hartley's call to action was as dramatic as in any fictional Quest story:

> One afternoon, I was in a pub with some friends when I thought I heard a stranger utter my name behind me. Whether real or not, it was the start of something. What began as a sense of unease turned within minutes to paranoia. I was sure that the other people in the pub were talking about me. I felt they wanted to kill me, and somehow I knew they had brought tools (chisels and sharpened screwdrivers) with them to do it.
>
> I tried to behave normally, but I was very uncomfortable, I kept shifting around, and my friends agreed to go to a different pub. But the fear continued to build, and I decided I needed to get far away.

The familiar, yet now strange, world that Hartley felt himself in certainly had the Quest ingredient of otherworldliness:

> Reality had taken on a much more urgent, florid vibe. Music was thumping everywhere I went. The city lights were flashing and pulsating.

There was even the sense, not of the supernatural per se – Hartley told me he is not a religious believer – but of that quasi-religious feeling of being an important part of something much bigger; of a set-up, Hartley told me, that required the participation of almost the entire city:

> I wasn't really seeing or hearing things, it was just that almost everything around me seemed to be endowed with a special

significance, and it all related to me! Every traffic signal, every flickering streetlight was a sign meant just for me.

Hartley describes feeling as though he were the protagonist of a real-life spy scenario, with a secret mission. Everything he saw and heard was a coded message. He describes hearing the line 'I met him at the candy store' from the Shangri-Las' song 'Leader of the Pack' in a real-life candy store, and taking it to be 'an amazing coincidence, laden with meaning'; only later realizing that the store's playlist consisted entirely of songs that mention candy. At the start of Act 2, as he put it, Hartley actually snuck into a courtroom and rifled through legal files, which he felt sure would hold – in coded form – the key to his mission.

But, they didn't. Hartley's Quest story just . . . petered out. As we chatted on Zoom, now more than thirty years after the event, what really came across was his – at the time – deflation at the way that his would-be Quest story failed to resolve itself. 'I was literally expecting to get given a quest or a mission,' he told me. 'I was looking at those papers in the courtroom looking to see a mission. And I was disappointed because I couldn't find it. It was like a plot hangover. James Bond on his day off. It's not happening'. The early ingredients were there – the call to action, the monsters (the guys in the pub), but the final ordeal never came.

It's interesting – and I think important – that the young Hartley experienced the curtailing of his Quest not as relief, but frustration. 'I had a feeling that this quest isn't emerging,' he told me. 'Where's the ring? Where's the volcano? It's the sense that it wasn't coalescing into a plot that was troubling and disappointing. It was the most boring and unresolved plot'. This only seems to confirm our need for masterplots. Even when experiencing psychosis, our need for a satisfying narrative – a story that follows one or other of our masterplot recipes – is always there. When I put this to Hartley, he not only agreed, but put it better than I ever could have done:

> We're all constructing a narrative, and trying to build a plot that makes sense. Everyone's trying to make their life conform to a masterplot all the time. And psychosis forces you to come up with a new plot, and it makes it more obvious that the need for narrative is there.

He then made a point that's so brilliant and insightful, I'm kind of jealous that I didn't come up with it myself. Why is it so annoying, he wondered out loud, when – as in the famous shower scene from *Dallas* – we're told that the events we've just seen were all a dream or a fabrication? I agreed, mentioning *Life of Pi*, which comes to much the same conclusion (or does it?!). Why do we react this way? We knew all along that none of this happened anyway; that we were watching a work of fiction. What does it matter – what does it even *mean*? – to be told that these fictional events 'didn't *actually* happen'?

The reason, Hartley suggested, is that when we watch a movie, we make a tacit agreement with the director to suspend disbelief; to commit wholeheartedly to the narrative. If we find out at the end that their commitment was half-hearted, we feel cheated. It's the same, he continued, for the aborted Quest in which he found himself. Having set itself up for a quest – as a result of all the Quest stories it had imbibed over the years – Hartley's brain felt short-changed when it ended up unable to provide itself with one. It was quite willing to suspend disbelief, to weave together just about any number of half-suitable events into a final ordeal; the universe just stubbornly failed to supply any.

Paradoxically, then, one of the strongest pieces of evidence for the fundamental importance of masterplots in understanding human behaviour is not the fact that the brain will always find a way to mix events together to create a coherent narrative – because it won't – but the fact that when it fails to do so, it reacts by generating a feeling of disappointment, of having let its owner down.

Masterplots aren't just a nicety; they're a compulsion.

HAPPY ENDINGS

In this chapter, we've explored the Quest masterplot, with its key ingredients of a call to action, a final ordeal and a life-renewing goal; and of course, the quest itself – to somewhere otherworldly, beset by monsters, the supernatural and great temptations, where our hero is caught between a rock and a hard place, before eventually making it home with the help of travelling companions and local helpers. We've seen how this masterplot has shaped works of fiction from *The Odyssey* and *The Pilgrim's Progress* to *The Super Mario Bros. Movie* and *Barbie*.

We've seen, too, how the Quest masterplot shapes real-life stories. Perhaps, in theory, Peter Moore *could* have learned his CEO skills without a quest to the USA, Karl Bushby *could* have become a science communicator without walking around most of the world, and I *could* have figured out my career choices without that trip to Russia. But we *wouldn't* have. It's the replacing of the old associations, the old predictions, with new ones – that otherworldliness – that changes everything, even when you're, geographically at least, back where you started. The Quest masterplot has remained all but unchanged since Homer's day, because it distils into its concentrated form our instinctive knowledge – which scientific evidence suggests is quite correct – that *place* matters: that moving to a different location changes us in some way, allowing us to break out of our ruts and to find that holy grail that we couldn't find at home, even if – as in all true Quest stories – it was there all along.

And for that reason, the Quest masterplot – perhaps more than any other in this book, although we will meet some strong contenders – has the power to shape human behaviour for the better. Yes, the Quest narrative can occasionally – as for Don Quixote and Tom Hartley – trick us by pulling together random events and misapprehensions into a seemingly coherent story: delusion, a core symptom of psychosis. But it can just as easily inspire, such as when Peter Moore returned home from the US with his life-renewing goal to – like the Ancient Greeks – conquer the (footballing) world; or when JFK vowed to put a man on the moon. If the tech bros are right, and we humans do eventually colonize new worlds, it will be the Quest masterplot that put us there.

3.
UNTANGLED

The King, expelled, for lust's embrace he sought,
His dalliances, the kingdom could not abide.
But Queen Rebecca, burdened by bitter thoughts,
In her heart, a plot brewed, country to divide.
To bring destruction to her own domain,
She chose a man incompetent, a fool.
Edward, unfamiliar with the land's reign,
Unversed in customs, naught but jest and drool.
Yet Edward's heart, with kindness ever blessed,
He shared his joy, his optimism grand,
And Nathan, lowest servant, oft caressed,
By Edward's warmth, their souls began to stand.
The people, moved by Edward's noble grace,
And Nathan's humble toil, without disdain,
Resolved to fight, their strength they would embrace,
For kingdom's sake, their efforts would remain.
Thus, Edward and brave Nathan took the lead,
Into the battle, against dreaded foe.
Though victory slipped away with grievous speed,
Their courage shone, their spirit all aglow.
Impressed was Queen Rebecca, touched, contrite,
Her diabolic plans she did disown.
In future days, she vowed with all her might,
To give her full support to Edward's throne.

        WILLIAM SHAKESPEARE – 'THE BALLAD OF TED LASSO'

THE STORIES OF YOUR LIFE

On the face of it, *Ted Lasso* – Apple TV's surprise hit of the pandemic, which tells the story of an American football coach who takes up a job at a football (soccer!) club in the UK – would seem to have little in common with 'The Adventure of Silver Blaze', the famous Sherlock Holmes mystery which tells the story of a murdered horse trainer. But dig a little deeper, and you find that both tales bear the same markings.

In 'Silver Blaze', Sherlock Holmes is summoned to the stables of the eponymous racing horse, where chaos – ahem – reigns. Silver Blaze is missing, and his trainer John Straker is lying dead on the moor, his head 'shattered by a savage blow'. In his pockets are a surgical knife – far too small and delicate to be useful as a weapon – and an invoice from a dressmaker. The price of the dress, twenty-two guineas (roughly £2,000 in today's money), is improbably high given the modest wages of a racehorse trainer, and indeed the invoice is made out to somebody else (one 'William Derbyshire'). The police have arrested a bookmaker who had been snooping around, chatting to the maid and the stable boy, but the case doesn't quite add up. OK, it's not hard to imagine that the bookmaker might have had a motive for horsing around with the outcome of a race, but if he had planned to kill the horse, why didn't he just do it there and then in the stable, rather than leading it off somewhere? And, if he really had murderous intentions, why would he make himself known to both the maid and the stable boy? The stable boy, by the way, was drugged, with opium mixed into his spicy mutton curry, but everyone else who ate the same meal remains mysteriously unaffected. Oh, and three sheep kept at the same stables recently went lame.

Before we go any further, let's give you the chance (if you don't know the story already) to play detective. Everything you need to solve the case is in the summary above. Whodunnit?

Did you get it yet? If not, don't feel bad, even Sherlock Holmes himself needed one more clue: one of the most famous clues in literary history, and one that would later lend its name to a best-selling novel:

> Inspector Gregory: Is there any point to which you would wish to draw my attention?
>
> Holmes: To the curious incident of the dog in the night-time.
>
> Inspector Gregory: The dog did nothing in the night-time.
>
> Holmes: That was the curious incident.

The dog didn't bark because it *knew* the person who led Silver Blaze out of the stable: the horse's trainer, John Straker. That's right, there never was a 'William Derbyshire'; this was a pseudonym that Straker used when he was carrying on an affair with his mistress (who demanded fancy clothes far beyond the means of a racehorse trainer). The surgical knife was found in Straker's pocket because his plan wasn't to *kill* Silver Blaze, just to give him an injury that would be small enough to be undetectable, but serious enough to hobble his chances in the big race. Straker was struggling to keep his mistress in £2,000 dresses, you see, and it was *him* who was looking to cash in by betting against Silver Blaze in the Wessex Cup, not the innocent bookmaker. This plan would likely have worked (particularly given Straker's trial runs with the sheep) had it not been foiled by Silver Blaze himself who – apparently spooked by his trainer's odd behaviour – kicked him in the head. Oh, and the mutton curry? Chosen by Straker himself as the only meal that would be spicy enough to hide the taste of the opium that he slipped into the stable boy's portion.

What happened to Silver Blaze? He was found wandering the moors by a neighbour, Silas Brown, who just so happened to be the trainer of Desborough, the second favourite in the Wessex Cup race. Silas hid the horse in his own stables, disguising him by dying the distinctive silver blaze on his forehead. Silver Blaze, now reinstalled as favourite, wins the big race; a great result for Sherlock who (at least, we are led to infer) backed Silver Blaze when he was still 'missing', at the generous odds of 15:1.

THE MASTERPLOT RECIPE

Seemingly tiny variations on a recipe, even when the key ingredients are the same, can lead to a very different end product, for culinary and masterplot recipes alike. A pinch of salt here, a teaspoon of sugar there, and the overall flavour is transformed. For no masterplot is this truer than the Untangled recipe, in which the same key ingredients can be used to create a comedy, a mystery or a thriller.

The first ingredient – and the one that is most key to the Untangled masterplot recipe is **the battle between light and dark**. At the start of *Ted Lasso*, the haughty owner of AFC Richmond, Rebecca, is the perfect embodiment of the dark. She is bitter from her divorce, jealous of the high-life enjoyed by the ex-husband from whom she inherited the club, twisted – wishing failure upon her own team – serious, unfriendly, stuck up and – yes – British. *Very* British. Ted is the perfect embodiment of the light: generous to a fault (he brings Rebecca homemade biscuits, even though she's unfailingly mean to him), optimistic, happy-go-lucky, a Dad-joker ('We're gonna call this drill The Exorcist, 'cause it's all about controlling possession'), gregarious, egalitarian and – above all – American. *Very* American.

An Untangled story is not – let's be clear – a battle between good and evil. In Monster (Chapter 5), the baddie is the embodiment of pure evil. In Untangled, the baddie is a pantomime villain. Yes, we boo them when they come onstage, and, yes, they have some dastardly scheme or other – hobbling their own team or their own horse – but they're not out to kill anyone. When **light eventually triumphs over dark** – the key ingredient at the end of an Untangled story – the villain is usually given a chance to repent and make amends. They might do a bit of light jail time, but they will certainly not – unlike Monster's monster – be slaughtered by the main protagonist. Similarly, the goodie in an Untangled story is not a po-faced embodiment of pure virtue, but a bit of a character. Ted is an unashamed goofball, and Sherlock isn't above using insider knowledge to line his pockets with a spot of illegal gambling.

In the Ancient Greek comedy *Lysistrata*, the battle is between the war-loving men of the town and their down-to-earth wives. In *A Midsummer Night's Dream*, it's between Egeus – the overbearing father who refuses to let his daughter marry the man she loves – and the fairies, led by their king, Oberon. In Netflix's *Beef*, it's between the overworked and overstressed business owner Amy Lau, and the underemployed contractor Danny Cho. Sometimes, like for Tom Hanks in *Big* or Bill Murray in *Groundhog Day*, the battle is fought between two sides of the same person: the dark, cynical adult and the light, fun-loving child. But *good and evil*? No.

As its name implies, though, an Untangled story isn't just any old victory of light over dark; that victory has to come about in a specific way. The second key ingredient to be added to the mix – right at the start of the recipe – is an **initial state of chaos**. As a result of a messy divorce, Rebecca has assumed control of her ex-husband's beloved football club, which she secretly plans to ruin by appointing a hapless American. Sherlock is faced with a missing horse, a dead trainer, a flimsy surgical knife, an invoice for a fancy dress, a spicy curry and some lame sheep.

But, again, it's not just any old state of chaos. This state of chaos has a particular flavour that comes from one – or more often both – of two further key ingredients. The first is **concealed or mistaken identity**. At least one of the major players is not who they seem to be. Perhaps they just hide their true intentions. Perhaps they literally put on a disguise. Perhaps they overhear something they weren't meant to know and have to feign ignorance. Perhaps they are a long-lost sibling or parent. Perhaps, in the most convoluted plots, all of these things; because the more tangled the ball of string, the greater the satisfaction you feel when you finally untangle it. 'Silver Blaze' has three cases of concealed or mistaken identity: the police have arrested an innocent bookmaker; John Straker disguises himself as 'William Derbyshire'; and Silas Brown disguises Silver Blaze as a run-of-the-mill horse. In *Ted Lasso*, Rebecca hides her true intentions from Ted, who clearly believes that the two share a common goal in the success of their team. *Ted Lasso*'s subplot combines concealed or mistaken identity with the second state-of-chaos ingredient – the **bizarre love triangle**. Jamie's girlfriend Keeley is photographed apparently (but not actually) kissing Ted (the mistaken identity), and then *actually* kissing Jamie's teammate Roy (the bizarre love triangle). If it's not done carefully, this bizarre love triangle element can feel tacked on, particularly if it's confined to a subplot that has little to do with the main story. But if it's done well, it all adds pleasingly to the overall flavour, since *maximum* chaos ensues when the concealed or mistaken identity and bizarre love triangle elements are combined and remixed (though sometimes ad nauseam).

Shakespeare, of course, is the master of the bizarre love

triangle. Take, for example, the plot of *Twelfth Night*, considered by many to be his best comedy. Viola falls in love with Orsino, who is in love with Olivia. But, oh no, Olivia has fallen for Cesario... who is actually Viola, disguised as a man. Meanwhile, in a side plot, Maria writes a prank letter to Malvolio – supposedly from Olivia – convincing him that Olivia is actually in love with *him*. Viola proposes to 'Cesario', who accepts, and they are married – but it turns out it wasn't Cesario/Viola at all, but Sebastian, Viola's twin brother.

If you're not following all of this, it doesn't matter: the key ingredient that comes right at the end – and that is 100 per cent absolutely non-negotiable – is the final **untangling**. Via a series of improbable discoveries, revelations and coincidences, everything comes out, the knots are untangled and everyone is free to win the horse race, football tournament or spouse that they were clearly supposed to win all along. As with the other key ingredients of this recipe, the untangling can play out in different ways, depending on the overall flavour that the writer is aiming for. In a knockabout comedy, the untangling is played for laughs, with a nod and a wink to the audience about how improbable and coincidental it all was (there's often an element of this in the Sherlock Holmes stories, despite their more serious subject matter). In a dark comedy – like *The Office* (the UK version) – the mood might be sad or poignant: Dawn and Tim finally kiss, but surrounded by interrupting colleagues, and leave together, but awkwardly and uncertainly. In a thriller, like *Paycheck* or *The Adjustment Bureau* (both based on short stories by Philip K. Dick) the improbable untangling is presented with a straight face: it's a *happy* ending, but not a funny one. Unless, of course, it's a comedy thriller like *The Sting* or a not-quite-a-comedy-but-definitely-knowingly-silly James Bond film.

But the untangling – like the state of chaos – can't just happen any old how. It must come about via the final ingredient: **help from an unlikely source.** In Shakespearean comedies, this is typically someone from the 'lower orders' of society (or, at least, someone disguised as one): a servant, a pauper, a jester, a child or – as in *A Midsummer Night's Dream* – a supernatural figure. In *Ted Lasso*, it's Nate: the team's kitman, caretaker and general dogsbody, who is at first bullied by some of the players, but who – in the final episode of Season 1 – is promoted to coach. Sometimes, particularly in the world of Disney, this unlikely source of help is a wisecracking animal sidekick. The name of this recipe is a nod to Disney's Rapunzel movie, *Tangled*, in which Pascal – Rapunzel's pet chameleon – trips up Rapunzel's nemesis and captor, Mother Gothel, sending her flying out of the window (in accordance with the Untangled recipe, she isn't killed as such; she just turns into dust). In 'Silver Blaze', Conan Doyle went one better than a wisecracking animal sidekick, with a famously silent dog. The help-from-an-unlikely-source ingredient is the icing on the cake of the Untangled recipe, because it tops off the take-home-in-a-party-bag moral cake that we have been baking all along: the idea that order will be restored, and happiness achieved, only when we set aside stuffy considerations of social hierarchy, power and money, and learn what really matters: spontaneity, following your dreams, love. Schmaltzy though it sounds, this lesson – and this masterplot – is one that we would often do well to follow in real life . . .

STRANGER THAN FICTION

In the summer of 1991, when I was thirteen, I owned just two pieces of recorded music. One was a cassette tape (ask your dad) that I'd purchased from Woolworths (ask your grandad) on the high street of the small town where I grew up in rural Suffolk.* The other was a record: a seven-inch single that I'd purchased from a local car boot sale. The tape – don't judge me – was Bryan Adams's single '(Everything I Do) I Do It for You' which, to this day, holds the record for the longest consecutive run at the top of the UK charts (sixteen weeks!). The record – fine, judge me – was a song called 'Ipswich, Ipswich (Get that Goal)', a novelty song released to celebrate my football team, Ipswich Town, getting to the final of the FA Cup in 1978 (which they actually won, beating Arsenal 1–0 in the final). Fear not, you can find this masterpiece on YouTube and Spotify.

Coincidentally, although I was completely oblivious to the fact at the time, both songs were written by the same person: Robert John Lange, a German-(via Zambia)-South African, known to his friends and enemies alike as 'Mutt'. Flush from his success with the Ipswich Town 1978 FA Cup Final squad, Lange went on to produce AC/DC's *Highway to Hell* (1979) and *Back in Black* (1980) and four Def Leppard albums. It was therefore something of a surprise when, in the early 1990s, he became such a fan of the 'Queen of Country Pop', Shania Twain, that he called her out of the blue to express his admiration and – not long after – to propose marriage.

When the happy couple moved to Switzerland with their young son, Shania hired a local woman named Marie-Anne Thiébaud to be her translator, assistant, estate manager, general dogsbody and – soon – best friend. Ah, yes, the servant girl; albeit a high-class one, wife of one Fréd Thiébaud, an executive at the Swiss chocolate and Nespresso giant, Nestlé. Well, you

---

* Woodbridge, if you're asking. Home to Charlie Simpson of the band Busted, and just down the road from Ed Sheeran's hometown of Framlingham. Hey, I never said it was cool.

can guess what happened next. Mutt wasted no time in getting his paws all over Marie-Anne. Shania had her suspicions that already twice divorced Lange was up to no good and confided – in true Shakespearean fashion – in her best friend and servant, Marie-Anne. Marie-Anne did her best to reassure her best friend and boss, apparently so successfully that Shania felt 'foolish' for even suggesting it. Later, her suspicions grew, and she confronted Marie-Anne more directly, but was again fobbed off: 'I'm heartbroken that you would even think that I was hiding something from you,' said Marie-Anne.[1] In fact, it wasn't until after Mutt had asked Shania for a divorce that she found out the truth. Fréd Thiébaud found hotel receipts and sexy lingerie in a suitcase that his wife had taken on a trip to spend some time 'alone', and told Shania what was going on.

'I rejected it initially,' Shania said later. 'But I couldn't control Fréd's love for me and how easy he is to love.'[2] That's right, Shania turned to Fréd for support – after all, he'd been through exactly the same thing – and one thing led to another. On New Year's Day 2011, they got married, and settled in Switzerland. The swap was complete. And, at least at the time of writing, both couples are still together and going strong.

Although it can't have been much fun for Shania, this real-life tale has all of the key ingredients of the Untangled recipe. We have the **battle between light and dark** – trusting Shania with her chart pop versus devious, brooding rocker Mutt who is *Back in Black* on a *Highway to Hell*. At the same time, Mutt is hardly evil personified. He behaved unacceptably, yes, but in a way that is par for the course for a rich, entitled, male celebrity. He would probably characterize himself as a loveable rogue (he calls himself 'Mutt', for God's sake). Accordingly, when light eventually triumphs over dark, Mutt isn't imprisoned or killed. He just slinks off with his tail (and everything else) between his legs, presumably vowing to behave himself from now on (a vow that – as far as we know – he has kept).

We start with the obligatory **initial state of chaos** characterized by **concealed or mistaken identity** and a **bizarre love triangle**. Mutt, an unreformed bad boy of rock, is playing the parts of pop producer and loyal husband. But he is in love with – or, at least, shagging – Marie-Anne. Marie-Anne, for her part, is playing the role of loyal best friend and confidante to Shania. We don't know, of course, whether anything had yet sparked between Shania and Fréd, but they're both very easy on the eye – a fact that each of them can hardly have failed to notice. Then, of course, it all comes out in the final **untangling**, whereby everyone ends up with the 'right' partner, via help from an unlikely source: someone – in true Shakespearean fashion – from the 'lower orders' of society. In Mutt's case, this was a literal servant (albeit an extremely highly paid one); in Shania's case . . . OK, a Nestlé executive is hardly slumming it, but it's certainly several steps down for an artist who is a fixture in the Top 10 of wealthiest female musicians (usually somewhere in the middle with Taylor Swift and Jennifer Lopez), with a net worth of $400 million. That's a lot of KitKats.

UNDER THE INFLUENCE

On a micro level, very little of Shania's story will resonate with us mere mortals, and I certainly wouldn't suggest that you embark on an affair with your partner's admin assistant. But if we zoom out and think about how this story embodies the overriding, unifying theme of the Untangling masterplot, we can all learn something. Its essence, remember, is the triumph of light over dark; the silly over the serious; the heart over the head;

the fun-loving child over the cynical adult. Shania found happiness when she stepped outside of the expectations that go along with her 'day job' – specifically, being married to a fellow multi-millionaire music heavyweight – and, yes, I make no apology for the cliché here, following her heart.

When it's viewed on this level, we can see that Shania's story is one that is echoed every day by people who took the plunge and did likewise. *The Guardian* even has a regular column called 'A new start after 60'; scrolling through the titles is like leafing through a screenwriter's 'ideas' notebook:

I became a powerlifter at 71 – and I've never felt so good about myself

I was unfit and prediabetic. A month tracking turtles changed everything

I embraced being single and became an international pet-sitter

I'm gay but had done nothing for LGBTQ+ people. So I used my pension to launch a lottery

I quit as a CEO and found my dream job as a truck driver

I found love after a painful divorce – and we moved to an uninhabited island

I became a naturist at 75 – and it felt like freedom

Sepsis almost killed me – but I survived to invent a new sport

Is there a single one of those that wouldn't make a fantastic feelgood Untangled movie? See, you don't have to swap partners with the man who wrote the 1978 Ipswich Town FA Cup Final song to live out your own comedy; just set aside the serious, and embrace the frivolous.

THE SCIENCE BEHIND THE STORY

To understand just how and why the Untangled masterplot works, we're going to have to delve a little deeper into an idea that we first met in Chapter 1: that of the brain as prediction machine.

To recap, the Untangled recipe consists of an initial state of chaos – characterized by concealed or mistaken identity and/or a bizarre love triangle – followed by resolution of that chaos – the triumph of light over dark – via help from the unlikeliest of sources. Why does that recipe give a final product that tastes particularly good?[3]

It helps if we start by asking why other things – like pizza, chocolate or sex – taste good. The reason, as most people know, lies with evolution. Mother Nature – that is, natural selection – decreed that humans (like other animals) should take pleasure from calorie-dense foods and mating with other humans, because if we didn't, we would have gone extinct long ago. Now, modern society has given us tools to 'trick' Mother Nature and enjoy these pleasures divorced from their original evolutionary benefits. Thanks to the technological advances of processed cheese and refined sugar, we can enjoy pizza and doughnuts, even though they add little to our chances of reproductive success (if anything, they probably have a negative effect). Similarly, thanks to the technologies of contraception and pornography, we can enjoy sex (or, at least, a simulation of it) without reproduction. It's just the same – so the theory goes – for stories that follow the Untangled masterplot. Modern technologies – novels, plays, movies, TV series and so on – allow us to enjoy the pleasure divorced from the original evolutionary benefits.

Just what are (or were) these evolutionary benefits? Essentially, clearing out the trash of our thought processes. As we saw in the first chapter, our brains are continually making predictions about the world around us, predictions that are updated millisecond to millisecond. If they weren't, we'd struggle with everyday tasks such as walking up stairs or holding a conversation. But because those predictions must be made very quickly, and based on incomplete or inaccurate information, they are often incorrect. These incorrect predictions need to be quickly identified and thrown out, before they seep into our long-term memories, our general knowledge about the world, and, like rotten apples, spoil the whole barrel. The problem is, expunging these incorrect predictions is boring, laborious work: who *likes* taking out the trash?

Mother Nature therefore hit upon an ingenious solution: rewarding the identification and expulsion of incorrect predictions with a mild hit of pleasure. In the context of a comedy, we interpret this pleasure as 'humour'. In the context of a mystery, we interpret this pleasure as . . . well, English doesn't have quite the right word for it but it's something like an 'epiphany', a 'revelation', an 'aha!' moment. The point is, it's pleasurable: nature's reward for expunging our incorrect predictions.

Although this theory was originally proposed to explain humour,[4] it seems to me that it applies to all stories that follow the Untangled masterplot, comedies and mysteries alike. But, in the interests of starting on the theory's home turf, let's reverse-engineer a famous joke by the British comedian Bob Monkhouse.

> I want to die peacefully in my sleep like my father; not screaming in terror like his passengers.

The humour comes because we make an incorrect prediction – 'His father died in bed at home' – which is then jarringly disconfirmed – 'Ah, no, he was driving a bus!' Or take my own favourite-ever comedy moment from *The Simpsons*. Krusty the Clown is having trouble in his relationship with his (until recently) long-lost daughter, and the Simpson family are doing their best to help. Marge suggests that Krusty's daughter may just need time to adjust, at which point Homer stands up, puts his hand across his chest and, with as much gravitas as he can muster, asks, 'Marge, may I play Devil's Advocate for a moment?' Marge agrees, and we cut to a scene of Homer in the Kwik-E-Mart playing a pinball machine called Devil's Advocate; the machine's graphics show the Devil wearing a lawyer's sharp suit.

The joke works because we commit hook, line and sinker to the prediction that Homer is about to make some insightful suggestion that will ultimately help Krusty to reconcile with his daughter (what else could 'May I play Devil's Advocate?' possibly mean?), which is then annihilated – we couldn't have been more wrong. Crucially, the more wrong we are, the greater the pleasure hit when we're corrected.

Now we're getting to the crux of the Untangled masterplot. The reason for all of the **initial chaos** – the **concealed or mistaken identity**, the **bizarre love triangle** – is to lead us in completely the wrong direction. There's no way, in *Ted Lasso*, that Roy will get together with Keeley when she's currently with Jamie, and has been spotted (apparently) kissing Ted. There's no way, in *The Office*, that Dawn will get together with Tim when she's in a taxi to the airport with her fiancé. There's no way that Silver Blaze's own trainer could be responsible for the horse's disappearance, when he has no apparent motive and – in a classic piece of Conan Doyle misdirection – is lying dead on the moor. Misdirection is key. A great comedy or mystery writer won't just successfully disguise the eventual outcome, but sow and nurture the seeds of incorrect predictions: the police have arrested the bookmaker . . . and he was snooping around the stables . . . and he does have a clear motive . . .

The greater the extent you commit to an incorrect prediction – so the theory goes – the greater the pleasure hit when this prediction is spectacularly trounced. This, then, is why the final triumph of light over dark has to be brought about via **help from an unlikely source**: it's the last thing you would have predicted. Who will turn AFC Richmond's season around: Ted? Quite likely. Rebecca? Maybe. Her ex-husband? Possibly. The kitman? No way! Who will give away what really happened to Silver Blaze? The bookmaker? Quite likely. Silas Brown, the neighbour? Maybe. The stable boy? Possibly. The stable *dog* . . .

PLOT TWISTED

What happens when you commit to a prediction that happens to be incorrect, and refuse, or actively rebuff, the efforts of the world and the people around you to expunge it? What happens when you reject the 'official' untangling of the clues in favour of your own, or one that you saw in a YouTube video? What happens is that you find yourself in the realms of conspiracy theory; of the Untangled recipe perverted.

Of course, it's not a conspiracy theory if it's true – and, over the years, more than a few 'conspiracy theories' have indeed turned out to be true.[5] The US Government, in particular, has done many of the things it's been accused of, including stealing body parts from recently deceased babies and children to test the effects of nuclear fallout; testing LSD and other hallucinogens on people without their knowledge (in the MK-Ultra programme); recruiting the Dalai Lama as a CIA agent (well, at least putting him on the payroll, to the tune of around $180k in the 1960s);

allowing 128 black men to die untreated in the Tuskegee Syphilis Experiment; spying on John Lennon; and hallucinating the second 'Gulf of Tonkin' incident that led to the Vietnam War. With a track record like this, we can certainly understand why some people were hesitant to trust the US Government when it urged its citizens – rightly – to get the new Covid-19 vaccines in late 2020.

Since all of these 'conspiracy theories' turned out to be true, how do we know that other popular 'conspiracy theories' aren't true too? This is tricky, as the very nature of conspiracy theories is that they inoculate themselves against debunkings by claiming that anyone who presents evidence against the 'conspiracy' must be in on it themselves, and probably in the pay of its shadowy overlords.

While we can never completely debunk a conspiracy theory – at least, to the satisfaction of its advocates – we can certainly show that this way of thinking doesn't add up, as was neatly demonstrated in a 2013 study that turned the tables on conspiracy theorists.[6] The methodology of the study was very simple: visitors to a diverse but 'broadly pro-science' climate change blog were given a survey in which they were asked to rate various statements that at least some of us (including me!) would call conspiracy theories: 'The moon landings were faked', 'Princess Diana was killed in a plot organized by members of the British Royal Family', 'Man-made climate change is a hoax', 'AIDS isn't caused by the HIV virus', 'Smoking doesn't cause cancer' and – the most dastardly conspiracy of all – 'New Coke [launched in 1985] was deliberately inferior and was part of a marketing ploy designed to boost sales when "Coca-Cola Classic" was reintroduced later'.

Before we get to the results, let's stop to think about what pattern we'd expect to see if one of these theories were true.

Let's say the moon landings really were faked. If this were the case, then we'd give a pat on the back — and a big apology — to those clever independent thinkers who managed to take a neutral scientific look at the available evidence and come to the right conclusion. There would be no reason for these clever independent thinkers to also reach the conclusion that smoking doesn't cause cancer, climate change is a hoax and Princess Diana was killed by the British Royal Family, since they judge each case on its merits, right?

Ah, no. Broadly speaking, the questionnaire respondents didn't pick and choose their conspiracies; they either believed in all of them or none of them. What this suggests — although of course you'll never persuade them — is that conspiracy theorists aren't independent free thinkers at all. In fact, they're 'sheeple' — even more so than the people they criticize. Regardless of the domain (climate change, aliens, fizzy drinks), regardless of the 'conspirators' (academics, the US Government, Coca-Cola executives) and regardless of the evidence, if someone says the official story isn't right, our conspiracy-thinkers will believe them.

The problem, then, is that conspiracy theorists are indiscriminate. They see *every* real-world story as following the Untangled masterplot, even when it's clear — to an objective observer — that it follows a different masterplot altogether. Take, for example, the moon landings; to my mind, the single best real-world example of a Quest. The Untangled version of the Apollo story has all the masterplot's key ingredients. At the start, all is chaos: humankind's greatest (would-be) achievement is a hoax, and — worse — almost everyone is taken in by it. Mistaken identity and disguised intentions? How long have you got? The short version is that the 'astronauts' were actors on a Hollywood sound stage, in a movie directed by Stanley Kubrick (supposedly, the reason why his 1968 movie *2001: A Space Odyssey* is so lifelike is that

Kubrick used the same techniques). The resolution of this chaos comes from the unlikeliest of sources, namely Bill Kaysing, an employee who worked for a company that helped to design Saturn V's engines, and who almost singlehandedly popularized the idea that the moon landings were faked:[7] the US flag is flying even though there's no wind on the moon (in fact, it was a special flag with a horizontal crosspiece); there are no stars in the photo (in fact, due to the short exposure needed to capture the astronauts and the moon itself); the shadows aren't right (in fact, light reflects off the moon's surface, as well as coming directly from the sun); you can't see Neil Armstrong's camera in the reflection of Buzz Aldrin's visor (in fact, it's built into his spacesuit) . . .

Of course, it's not just the moon landings. For someone who sees every major real-world event through the lens of the Untangled masterplot, the identity of someone (or something) is *always* being **concealed or mischaracterized** (and, if not because of an actual *romantic* **love triangle**, because of a love-in or unholy alliance between seemingly unlikely bedfellows). There is always some **chaos to be resolved,** and to be resolved with **help from an unlikely source,** such as the Belgian doctor Kris Van Kerckhoven, who speculated about a possible link between 5G masts and Covid-19,[8] or the anonymous internet commenter who claimed – with reference to the 9/11 attacks – that 'Jet fuel can't melt steel beams'. The unlikely-source part is crucial because, just as with intentionally fictional Untangled stories (as opposed to unintentionally fictional conspiracy theories), the more improbable the untangling, the greater the pay-off when it plays out.

Seriously, who falls for this stuff? The answer, a 2020 study suggests, is people who pervert or twist the Untangled masterplot by imposing it in situations where it just doesn't apply. In a study that must have been great fun to set up, psychologists (and professional sceptics) Gordon Pennycook and colleagues asked an AI to generate 'pseudo-profound bullshit' quotes,[9] like: 'Hidden meaning transforms unparalleled abstract beauty', 'Wholeness quiets infinite phenomena', 'Imagination is inside exponential space-time events', that kind of thing. The researchers found that people who rated these sentences as profound were more likely to agree with conspiracy-theory statements such as 'The power held by heads of state is second to that of small unknown groups who really control world politics', 'The rapid spread of certain viruses and/or diseases is the result of deliberate, concealed efforts of some organization' and 'Secret organizations communicate with extra-terrestrials, but keep this fact from the public'. A later study, also led by Pennycook, found that people who ascribe profundity to this 'pseudo-profound bullshit' are also more likely to accept fake news as true.[10] Together, the findings of these two studies suggest that, while we are all in thrall to narrative to some extent, some people just have a very low bar for interpreting the everyday chaos all around us as a series of clues that can unlock some vast mystery; in the terms – that is – of the Untangled masterplot.

HAPPY ENDINGS

Stories that follow the Untangled masterplot recipe, whatever their genre, are fundamentally a battle in which light triumphs over dark. We start with chaos: people or things aren't who or what they seem, are struggling with their true identity, or are in a relationship with the wrong person. But finally, via help from the most unlikely of sources, all of these twisted threads are untangled, and harmony is restored to the universe. The lesson that Untangled stories teach us, then, is that the path to happiness is rejecting the dark – social hierarchy, power, money, the conventional – and embracing the light – spontaneity, following your dreams, love, the route less travelled.

'Huge if true', as they say. But is it true? Can applying the Untangled masterplot to our own lives really serve as a catalyst for putting us on the road to success, happiness and a fulfilling life?

The answer is a cautious, 'Yes, if you're sensible about it.' Don't – for goodness' sake – give up your career to pursue your dream job: a recent survey found that just 4 per cent of adults managed to make it into the job they dreamed of as a child (though 'lawyer' and 'doctor' buck the trend somewhat, with 14 per cent and 10 per cent respectively achieving those rather prosaic dreams).[11] Certainly don't set your heart on becoming a professional YouTuber – the career goal of almost a third of eight to twelve-year-olds today.

In fact, it's best if your dreams are not work-related at all. A study conducted in Germany found that setting life goals based on career and financial success was generally associated with lower life satisfaction, presumably because most people fail to achieve them.[12] Those who set goals relating to friends, family, their social lives and getting involved with political causes were happier and more satisfied with their lives. A great rule of thumb – from the writer David Brooks – is to think about what you'd like to have written on your gravestone, or mentioned in your eulogy:[13] not 'in the top 1 per cent of earners' but 'a dad who always put his kids first'; not 'author of a hundred scientific papers' but 'fought tirelessly for the causes he believed in'. Think of those 'A new life at sixty' articles we met earlier: taking up powerlifting or pet-sitting, tracking turtles or trucking, inventing a new sport, launching a lottery; these aren't trivial goals, but they're ones that are achievable (and indeed, ones that their setters achieved). The Untangled masterplot can act as a catalyst for human progress, then, when we heed its lesson to prioritize the light over the dark, the silly over the serious, the heart over the head, and follow our (modest) dreams.

4.
ICARUS

When shall we three meet again?

My English teacher used to say that the first line in a film, novel or play sums up what the whole thing is about. I used to think this was his own brilliant insight, but as it turns out, it's a well-known device. The three main characters of *Macbeth* – Macbeth, Lady Macbeth and Banquo – are meeting at a crossroads in their lives, faced with impossible choices that will determine not only their own fates, but that of Scotland. What should they do? When, where and how will their paths intermingle? When shall we three meet again? Sometimes, the device is more up-front and explicit, as in quite possibly the best opening line ever, 'The past is a foreign country: they do things differently there' (from L. P. Hartley's 1953 novel *The Go-Between*).

It's a similar story for *Parasite* – at the time of writing, the only non-English-language film to win the Best Picture Oscar. 'Not you too,' says our hero Ki-Woo, 'upstairs neighbour finally locked up his Wi-Fi'. *Parasite* is about how the poor family in the basement can't afford the nice things enjoyed by those 'upstairs', and their efforts to instead acquire them by means that are perhaps not exactly *illegal* as such, but certainly shady, and therefore sure to end badly . . .

As we join the story, Ki-Woo and his family (the Kims) are unfulfilled and unhappy, scraping together a living by assembling pizza boxes in a basement flat, with a window that drunks use as a public toilet. *Parasite*'s call to action comes when Ki-Woo's friend mentions that he is leaving his job as an English tutor for the daughter of a wealthy family, the Parks, and suggests that Ki-Woo poses as a university student and takes over his job; which he does, re-branding himself as 'Kevin' (and having a fling with the girl he's tutoring). Flush with success, 'Kevin' suggests that his sister repeat the trick. 'Jessica' duly takes up the job as art therapist to the son of the Park family. Things are going very well; better, in fact, than Ki-Woo/Kevin could have hoped. Had he decided to leave it at that, he and his sister would almost certainly have got away with it indefinitely. The Parks had no qualms over their tutoring work, and no reason to suspect anything untoward.

But this isn't enough for Ki-Woo/Kevin – the poverty of his parents continues to eat away at him, and he has to act. Now things take a dark turn. Working with Ki-Woo/Kevin, Ki-Jung/Jessica frames the Parks' driver by leaving a pair of knickers in his car. He is dismissed, and the siblings' father, Mr Kim (though they keep his identity secret from the Parks) brought in to replace him. The long-standing housekeeper pays a similar price, in a fast-paced sixty-shot montage that brings the movie's first act to a shocking climax. As the housekeeper sits doing the family's accounts, Jessica sprinkles peach shavings onto the back of her neck, knowing that she has a severe allergy. Meanwhile, Mrs Park's new driver (Mr Kim, remember) is telling her that he overheard the housekeeper mention something about having tuberculosis, a serious and contagious disease. As Mrs Park enters the house, she sees her housekeeper coughing into the bin. Echoing Lady Macbeth holding up her (imaginarily) bloodstained hands, the grim-faced driver holds up a tissue stained with red sauce. Game, set and match. The housekeeper is dismissed and Mrs Kim takes her place (though, again, they keep her identity secret from the Parks). Finally, the whole Kim family are in the Parks' employ.

With the Parks away on a camping trip, the Kim family enjoy their new status as Lords of the Manor, partying in the Parks' fashionable modern mansion. But then Banquo's ghost shows up in the form of the sacked housekeeper, who reveals that her husband has been – and still is – living in a hidden basement. Hearing the Parks return early from camping, the family bundle the hermit husband and his ex-housekeeper wife into the basement, fatally wounding her in the process.

The climax of the film is a birthday party at the Parks' for the son of the family. Ki-Woo/Kevin plans to finish off what he started by killing the housekeeper's husband with a rock; but the husband turns the tables and smashes the rock onto Ki-Woo/Kevin's head. He survives, but his sister Ki-Jung/Jessica isn't so lucky – the housekeeper's husband stabs her with a kitchen knife in front of the horrified party guests. This doesn't go down well with her mother, Mrs Kim, who fatally stabs the housekeeper's husband with a barbecue skewer. Adding insult to barbecue-skewer injury, snobby Mr Park reacts disgustedly to the housekeeper's husband's 'lower class' odour. In a touching, if bloody, moment of class solidarity, Mr Kim stabs Mr Park in the chest, then flees – he really should have thought this through – into the basement. The movie ends with Ki-Jung/Jessica dead, Ki-Woo/Kevin back in the dingy flat with his mother, and her husband still trapped in the basement. Like the prophecy of *Macbeth*'s three witches, the prophecy of *Parasite*'s upstairs-neighbours' Wi-Fi turns out to have been not just correct, but self-fulfilling: the harder you try to get your hands on the upper classes' nice things, the more you're sowing the seeds of your own destruction.

THE MASTERPLOT RECIPE

I've called this masterplot Icarus – instead of the more-traditional Tragedy – because, in everyday conversation, the latter term has been diluted to mean nothing more than 'something really bad happening'. This distinction will become particularly important towards the end of the chapter when we explore some real-life events that are certainly tragedies, but that should not be misinterpreted as Icarus stories. Because an Icarus plot isn't just one in which bad things happen – although they always do – it's one in which bad things happen for a particular reason, and in a particular way. Indeed, while many of the masterplot recipes we'll meet in this book have a degree of flexibility, the five key ingredients of an Icarus story – **dissatisfaction**, **temptation**, **elation**, insation (OK, OK, **insatiability**) and **destruction** – must always be added in exactly that order.

The first key ingredient of the Icarus recipe to be added to the mix is **dissatisfaction**. Our hero, when we meet them at the start of the story, is restless, bored, unfulfilled, unhappy, stressed-out, worried. Ki-Woo is folding pizza boxes in a dingy basement. Macbeth, the relatively lowly Thane of Glamis, doesn't enjoy the status that he thinks his military victories should have earned him. Icarus, our eponymous tragic hero, along with his father Daedalus, is trapped in a labyrinth (a punishment meted out to Daedalus for helping Theseus escape from the same labyrinth after killing the famous half-man, half-bull Minotaur).

ICARUS

Next – and, yes, this must always be the second ingredient – comes the **temptation**. The anti-hero is presented with a possible way out of their dissatisfaction. 'Hey, Ki-Woo – why don't you pose as a university student and take over my job as an English tutor?' 'Hey, Macbeth, Thane of Glamis –' this is the witches talking – 'how about also being Thane of Cawdor, and then King of Scotland?' 'Hey, son,' says Daedalus, 'how about we strap on some wax-and-feather wings and bust out of this joint?'

As we'll see throughout this book, all masterplot recipes have some kind of *call-to-action* ingredient at about this point; that's part of the three-act structure we met in Chapter 1. But in the Icarus recipe, it has a dark flavour, which is what makes it a *temptation*. This dark flavour comes from three sub-ingredients that combine to make the temptation. The first is the *Achilles heel*.* The temptation prays on the defining character flaw of our anti-hero: Macbeth's ambition, Icarus's pride and vanity, Ki-Woo's resentment of his family's poverty. The second is *transgression*. In a Quest story, the thing that the hero is called on to do is good and virtuous. In an Icarus story, the thing that the anti-hero is called on to do flies in the face of convention, or morality, or the laws of nature. It's arrogant – Icarus snubbing his nose at the laws of nature by flying like a bird; it's at least morally questionable – Ki-Woo posing as a university student; often, it's bang out of order – Macbeth killing King Duncan and usurping his throne. The third sub-ingredient of the temptation is *dilemma*. In a Quest story the hero doesn't hem and haw: the

---

* If you don't know your Greek mythology, Achilles's mother dipped him in the River Styx to make him invulnerable. In a bit of an oversight, she held him by his heel, and didn't think to double-dip, meaning that his heel remained unprotected. An 'Achilles heel', then, is a weak spot, particularly in someone or something that – like Achilles – is otherwise invulnerable (like, for example, the thermal exhaust port on the Death Star).

call to action is irresistible, and they follow it right away. In an Icarus story, the hero is tortured by indecision. They know, on some level, that God or nature or fate has put that path before them because it targets their Achilles heel; they know that it's a transgression, and not really the right thing to do, and yet . . . and yet . . . Just as with everyday temptations, then ('Do you want some chocolate fudge cake?'), the Achilles heel ('Well, I do have a weakness for chocolate . . .'), the transgression ('but I'm on a diet . . .') and the dilemma ('so I really shouldn't') combine to create a temptation that – despite it all – the tempted was never going to resist.

It's important to point out here that the Icarus recipe can yield end products that have very different, almost contrasting, flavours. In the 'sweet' version of the Icarus story, the flavour here is pure Schadenfreude: the hero – or rather the anti-hero – gets what they deserve. *Macbeth* is a prime example here, along with – say – *The Wolf of Wall Street* or *The Bonfire of the Vanities* (Wolfe on Wall Street?). In the 'bitter', poignant version – *Romeo and Juliet, Titanic, Gone with the Wind, Brokeback Mountain* – we are rooting for the central character, and are horrified when the tragedy unfolds. Sometimes – *Parasite, Crime and Punishment,* the Icarus myth itself – it's a matter of perspective or interpretation. Is Ki-Woo a violent criminal or a victim of circumstances? (It depends, really, on your personal politics.) And, OK, so Icarus was a bit cocky, but did he really deserve to die? We can easily imagine a modern Icarus which is much more sympathetic to its central character, and his desire to soar with the birds (he was trapped in a labyrinth, OK?).

Crucially, though, the recipe ingredients are the same. Whether we are dealing with a sympathetic hero, an unsympathetic anti-hero, or something in between, this central character must encounter a **temptation** that targets their *Achilles heel*, requires some kind of *transgression*, and therefore places them in an uncertain *dilemma*. Romeo and Juliet, for example, share the Achilles heel of impulsiveness, must transgress the boundaries set up by their warring families, and find themselves in a dilemma as to what to do (played out, most famously, in Juliet's balcony soliloquy).

Inevitably, then, our (anti)hero succumbs to temptation. Next – remember, the order of these ingredients is non-negotiable – comes **elation**. Things aren't just going well; they're going much better than could have reasonably been expected (literally, 'elated' means 'raised above'), *and nobody suspects a thing*. Ki-Woo was only after a tutoring job; now he's screwing his client literally as well as metaphorically, and has got a job for his sister too. Macbeth not only kills Duncan, but looks on in delight as Duncan's sons flee the country, implicating them in the murder. Icarus not only escapes the labyrinth, but has a whale of a time soaring through the sunny Greek skies. Romeo and Juliet not only get together, but manage to find a friar who will marry them in secret.

If only our hero could quit while they were ahead. But in an Icarus story, they never do, thanks to the next ingredient to be added to our recipe: **insatiability**. At this point, the same character flaw that led to our anti-hero succumbing to temptation rears its head again, and is now the reason that – despite their early successes – they remain unsatisfied. It's not enough for Ki-Woo to have got jobs for himself and his sister; he wants to get the whole Kim family in on the scam. It's not enough for Macbeth to be King of Scotland; he wants a line of succession, too (the witches having prophesied that it would be Banquo's descendants, not his, that would be future royalty). It's not enough for Icarus to have escaped the labyrinth and enjoyed a leisure flight over the Med; he wants to fly closer and closer to the sun. It's not enough for Romeo to have Juliet . . .

So, our anti-hero pushes their luck. Or, rather, pushes it even further. They were already going out on a bit of a limb, but now they're just taking the piss. Emboldened by having gotten away with their first transgression, they do the same thing again; only more so. Ki-Woo ups the ante from harmless minor deception to framing the Parks' innocent driver and housekeeper. Macbeth murders not only Banquo (which is a bit rich, though killing off rival claimants to the throne was fairly common practice at the time), but also the innocent wife and children of another potential rival, Macduff (which is pure brutality). Icarus, despite his father's warnings, flies higher and higher. Romeo kills Tybalt.

Deaths – often murders – are common at this stage; to the extent that **death of supporting characters** qualifies as an ingredient, if only an optional one (in the Greek myth, for example, nobody dies except Icarus himself). In *The Seven Basic Plots*, Christopher Booker identifies four archetypical characters who are ripe for the chopping block: the good old man (Duncan; Mr Park), the rival or shadow (Banquo; the housekeeper's husband), the innocent child (Macduff's children; the Park daughter, who doesn't actually die, but witnesses the death of her father, and is just about the only sympathetic character in the film) and the temptress (Lady Macbeth; and possibly to some extent the Park daughter). Why these four? Each has a symbolic role: the destruction of authority, peers, innocence and love, respectively.

The final key ingredient to be added to the Icarus masterplot recipe is the **destruction** of the central character. Usually, as for Macbeth and Icarus, Romeo and Juliet, they are killed; and in fairly grizzly circumstances. Occasionally, they are allowed to remain alive, but only if they are so broken that death would arguably be preferable: *Parasite*'s Ki-Woo must live with having caused the death of his sister and the virtual imprisonment of his father.* Importantly, the circumstances of this death or destruction aren't random, but are brought about by the (anti) hero's own actions. We all knew all along – the protagonist included – that they were sowing the seeds of their own destruction. Macbeth is killed by Macduff, with the help of an army that came together precisely to overthrow the tyrant that Macbeth had become. Icarus is ultimately killed by height: exactly what he had sought by flying close to the sun. Ki-Woo is knocked out, and his sister killed, by someone from the lower classes who'd tricked their way into the Parks' house – exactly the trick that Ki-Woo himself had been trying to pull off.

But that's just in the movies, right? There's no reason real-life should be so poetic . . .

STRANGER THAN FICTION

If, like me, you live in the UK, it won't have escaped your notice that things are – well – a bit shit, right now. For starters, postal workers, train staff and barristers have all been on strike, meaning that – even if they received the summons and made their own way to court – victims and defendants in criminal cases found their hearings cancelled. The price of electricity and gas is at an

---

* An alternative reading of *Parasite* has Ki-Jung/Jessica as the main character. She *does* end up dead in grizzly circumstances and, having framed the Parks' old driver with her knickers, is arguably more of a Macbeth/Icarus figure than her brother. On the other hand, the story is generally told from Ki-Woo's perspective, and as someone who obviously cares deeply about his family, perhaps the death of his sister and the imprisonment of his father represent a greater destruction to Ki-Woo than would his own death.

average of around £3,000 per household per year; down from a high of £5,000, but still *three times* the highest price found in mainland Europe.[1] Thousands of businesses have closed as increasing rents and energy costs turned profitable firms into loss-making ones overnight. Queues at ports and airports mean that cancelled holidays and empty supermarket shelves are routine. Meanwhile, homegrown crops rot in the fields as farmers can't find staff to pick them. Inflation is currently around 5 per cent, although in 2022 it topped 11 per cent.[2] To cap it all – and this is not a metaphor – raw sewage is pouring into the sea, the government having voted down an amendment which would have legally required water companies to reduce this discharge.

It would be an exaggeration to say that all of this was brought about by one man, but not much.

The first thing you need to know about former UK Prime Minister Boris Johnson is that 'Boris' isn't his actual name. Well, not really. His full name is Alexander Boris de Pfeffel Johnson, and his friends and family call him Alex. 'Boris' is a character that Alex created as part of an effort to fulfil his childhood dream of becoming 'World King' – no, that's not an exaggeration. Because with 'Boris' Johnson, everything is an act. He's not the bumbling buffoon that he makes out to be in interviews. As a former journalist, Alex knows how to put together a story, and Boris is its central character. When Boris says or does absolutely *anything*, the *only* consideration is how it feeds into this narrative. Not whether it is best for the country, best for his own party (the Conservatives, roughly the UK equivalent of the US Republicans) or even – in the long run – best for Alex Johnson. Not whether what he's saying is true, or plausible, or even possible. The only consideration is whether it sounds good to his supporters, whether it adds to the glory of our (anti)hero, Boris.

Just as *Macbeth* starts with a prophesy from the three witches,

ICARUS

the story of Boris Johnson starts with a prophesy from the man himself. When he was just a lowly Member of Parliament, Boris set out his plan for world domination in a thinly fictionalized political novel called *Seventy-Two Virgins*. As you might expect, Boris is no Tolstoy. His idea of sophisticated humour is making the jumpy tow-truck driver character Serbian, just so he can call him Dragan Panic. Geddit?! Great literature it is not, but as a self-fulfilling prophecy of Boris's own Icarus story, it is on the nose.

'It was, in the end, your friends who did you in,' muses the Boris character, Roger Barlow.

'And quite right, too. That was what friends were for.'

Our story starts in 2012 at the London Olympics. Great Britain has just won its first gold medals – Helen Glover and Heather Stanning in the rowing – and is, for once, feeling pretty good about itself. But who's that, dangling from a sagging zipwire in the cloudy London skies, feebly waving a pair of Union Jacks? Yes, it's 'Boris' Johnson. As he waits to be rescued, Johnson daydreams in flashback . . .

A rowdy newsroom in Brussels, Belgium. Background noise of shouted telephone conversations in French and German. At a desk sits a cherubic red-faced young man with a mop of blonde hair, wearing a tweed sports jacket, a white shirt and a red tie with horizontal gold stripes. He holds a large glass of red wine. The empty bottle sits on the table. He shouts into a telephone.

```
BORIS: Of course it's true. I mean, they're
    not actually . . . But of course . . .
    Well, anyway, this one's a great one.
    This'll make a great headline – how
    about . . . BARMY BRUSSELS BENDY BANANA
    SHAKE-UP? Or don't we have shakes in
```

> the UK? How about BANANA BALLS UP? Bit mixed metaphors? Anyway, fuck it, that bit's up to you. Ready? Seriously, you're gonna love this. Ready? Barmy Brussels bureaucrats were up to their usual monkey business yesterday, when they published plans banning – put that in capitals – BANNING the sale of bananas that are – capitals – TOO CURVY. Exclamation mark. The freaky fruiterers; that's F, R, U . . .

OK, so Boris didn't actually write the infamous Bendy Bananas story, but as *The Telegraph*'s man in Brussels – the headquarters of the EU – he phoned in many, many stories along those lines. He also mentioned the supposed banana ban frequently during his subsequent Brexit campaign. (The story does, incidentally, contain a grain of truth, but not much more. The EU law states that only bananas that are 'free from malformation or abnormal curvature of the fingers' can be sold as 'extra' grade. Those with defects can be sold only as Class 1 or Class 2.)

The flashback speeds up into a quickly cut montage. We see Boris taking up editorship of the political magazine *The Spectator*, becoming elected as Member of Parliament, and presenting a satirical panel quiz, *Have I Got News For You?* for the BBC.

Boris snaps open his eyes on the zipwire and we are back to the present day. He's doing well enough, but has an air of **dissatisfaction** about him. 'Mayor of London' isn't a bad job title, but it has a bit of a 'Thane of Glamis' ring to it; it's hardly 'World King'. Worst of all, his old Eton and Oxford rival, David Cameron, is prime minister.

The **temptation** came when Cameron decided to hold a referendum on Brexit, a decision that he will probably regret every day for the rest of his life. It's not as if the British public were clamouring to leave the EU, or even to have a referendum. Cameron's Conservative Party were just losing a few votes to the UK Independence Party, who were demanding such a referendum. In the miscalculation of the century, Cameron thought he could put the issue to bed once and for all by holding the damned referendum, winning it comfortably for 'Remain', and thereby putting UKIP out of business. Well, as you may have heard, that's not quite how it turned out . . .

Why not? Thank our man Boris, who suddenly had a decision to make. Would he line up with the establishment, his own prime minister, and just about every ex-prime minister and world leader who spoke up on the issue, and back Remain? Or would he choose the path of chaos and back Leave? Well, we already know how Boris approaches these types of decisions. Siding with the establishment was hardly good for his brand. The public loved the iconoclast, the outsider, the bumbling idiot who does stupid things just for the sheer hell of it. And, make no mistake, as the son of an EU Commissioner, Boris knew that leaving the EU was stupid. Like Cameron, he probably thought there was no realistic possibility of it happening anyway. So why the hell not? Campaign for Leave, thumb your nose at the establishment – particularly Cameron, with whom Boris had long had a particular enmity – crack a few gags about the Barmy Brussels Bureaucrats and their bendy bananas, and build brand Boris.

Boris's Brexit temptation has all three of the sub-ingredients for temptation that we met above. First it targets his *Achilles heel*. No, not his hatred of the EU or of David Cameron, but his need to be liked. Boris is a *Pagliacco*, a sad clown whose buffoonery hides a deep sense of insecurity, perhaps stemming from his time at boarding school, where – in the 1970s – pupils were still regularly beaten by teachers; perhaps from his parents' divorce. You can see this need to be liked most clearly in interviews: Boris will never pass up the chance to say something that will get his audience to like him, to build his brand, no matter how apparently tenuous its relationship with the truth. Take the time a radio interviewer asked Boris what he did to relax. He could have easily batted away this softball with a nothing answer about reading, jogging or being too busy for relaxation. Instead, he span a bizarre yarn about building models of London buses out of old wine boxes, awkwardly segueing in a boast about how he had brought the real thing back to London's streets. Or take a 'What's in your shopping basket?' puff piece in *The Observer*. Why give a straight answer when you can act the clown? At night, 'a few lettuce leaves with tuna is enough'. In the morning 'I might even have some cold spaghetti or a chop or two, if there is some left over from my kids' meal the night before. There are often birthday cakes in the fridge, which I'll eat for breakfast.' In other words, 'I'm fun and relatable. Please like me.'

The second sub-ingredient of the temptation, remember, is *transgression*. As we've already seen, just about every grown-up in the room – including not just Boris's boss Cameron, but his own father, Stanley – was against Brexit. Backing the Leave campaign was arrogant; it flew in the face not just of convention, but of the political laws of gravity. It was transgressive in every sense of the word.

Accordingly, Boris faced the third sub-ingredient of the temptation: *dilemma*. Our anti-hero hems and haws before eventually choosing the dark path that preys on his character flaw. And so it was with Boris and Brexit. If the tragedy of BoJo was a work of fiction, any editor worth their salt would have objected to the following plot point as both trite and implausible, but it really happened: before deciding whether to back the Leave or Remain campaign, Boris wrote two newspaper columns – one backing each – and then decided which he liked more. If you don't believe me, just google 'Boris Johnson Pro-Remain article'; it remains freely available online.[3] First, he tugs on our heart strings:

> Shut your eyes. Hold your breath. Think of Britain. Think of the rest of the EU. Think of the future. Think of the desire of your children and your grandchildren to live and work in other European countries; to sell things there, to make friends and perhaps to find partners there. Ask yourself: despite all the defects and disappointments of this exercise – do you really, truly, definitely want Britain to pull out of the EU? Now?

Then he makes – as it would later turn out – an extremely prescient point about Vladimir Putin:

> And then there is the worry about Scotland, and the possibility that an English-only 'leave' vote could lead to the break-up of the union. There is the Putin factor: we don't want to do anything to encourage more shirtless swaggering from the Russian leader, not in the Middle East, not anywhere.

He ends with the economic argument:

> It is surely a boon for the world and for Europe that she [Britain] should be intimately engaged in the EU. This is a market on our doorstep, ready for further exploitation by British firms: the membership fee seems rather small for all that access. Why are we so determined to turn our back on it? Shouldn't our policy be like our policy on cake – pro having it and pro eating it? Pro Europe and pro the rest of the world?

Then he published the Leave column and backed Brexit. But he surely knew – like all Icaruses do – that he was taking the dark path. When the Leave campaign – against all predictions – won the vote, Boris appeared on TV looking, as one commentator put it, 'ashen-faced in victory'.[4] This was just a jolly jape; Brexit wasn't actually supposed to *happen*. That's why – when Cameron immediately resigned – Boris cannily gave the leadership contest a wide berth. He knew that the new Conservative party leader – who under UK law would become prime minister by default, no election needed – would face the impossible task of trying to deliver the undeliverable.

This unenviable task fell to an unpopular and – even by politicians' standards – unusually wooden colleague, Theresa May, who was hobbled not only by her woodenness and by systemic sexism, but also by the fact that – like most mainstream politicians – she had publicly backed the campaign to Remain. Boris sniped from the sidelines while May negotiated a settlement with the EU, one that she couldn't get her own party to back. Then, when she was forced to resign, Boris repackaged substantially the same deal as *Boris's Oven-Ready Brexit* and won the leadership election (and, later, a general election) by a landslide. Cue much eye-rolling from every woman who has had their idea enthusiastically stolen by a senior white male in a board meeting.

For Boris, cue **elation**. At this stage, remember, things are going much better for Leave voters' hero than anybody could have anticipated. Boris didn't just win the election, he smashed it, winning the biggest majority since Margaret Thatcher in the 1980s.*

So what did Boris do? Keep his nose clean, stay out of the

---

* OK, so his party won only 43.6 per cent of the votes; but under the UK electoral system (which shares similar deficiencies to that of the US), this translates into a huge eighty-seat majority in Parliament.

newspapers and use his huge majority to turn around a flagging, flailing Britain? Of *course* not. The next ingredient, remember, is **insatiability**. Having got away with his ludicrous scheme to back Brexit, kill the May Queen and usurp her throne, he naturally wanted to see how far he could push it. And the answer was 'an incredibly long way'. Here are just a few things that Boris did that didn't dent his sky-high poll ratings one iota:

- Advised the Queen to suspend Parliament for five weeks, meaning that Members of Parliament wouldn't get the chance to debate his Brexit deal (which Boris himself would later try to rip up). A court later decided that his advice was based on claims that were flawed and untrue.

- Failed to attend any of the five emergency meetings convened to plan the UK's response to the Covid pandemic and – at least according to one study – missed the opportunity to save 20,000 lives by beginning the first lockdown seven days earlier.[5] He also caught and spread Covid himself, boasting 'I was at a hospital where there were a few coronavirus patients and I shook hands with everybody.'[6]

- Attended a 'bring your own booze' party at his official residence, 10 Downing Street, in violation of his government's own lockdown rules. Overall, the police issued 126 fines for Downing Street lockdown breaches. Yet Johnson claimed in Parliament that 'All guidance was followed completely during [sic] Number 10.'* Ultimately, Johnson too was fined by the police; the first prime minister ruled to have broken the law while in office.

---

* The website https://boris-johnson-lies.com/johnson-in-parliament lists more than seventy lies to Parliament, though the convention is that even one is a resigning matter (indeed, MPs are prohibited from even *suggesting* that a colleague is lying in Parliament).

- Backed his special advisor, who seemingly broke lockdown rules by driving across the UK while Covid positive, saying that he had acted 'responsibly and legally and with integrity'.

- Breached electoral law by asking a friend, Lord Brownlow, to pay for the renovation of his Downing Street flat, including infamous rolls of gold wallpaper costing over £800 each. The electoral commission later ruled against Johnson on this issue, fining his party £17,000.

- Oversaw the purchase of over £4 billion worth of unsuitable Covid personal protective equipment (PPE) via a 'VIP lane' set up for MPs' contacts, an arrangement ultimately ruled unlawful by the High Court. The PPE was burned.

- Personally intervened – at least according to two whistle-blowers – in the evacuation of Afghanistan to ensure that ninety-four dogs and sixty-eight cats were prioritized, leaving behind many Afghans who had worked for the British Army, who were then at the mercy of the Taliban. Johnson called the claims 'total rhubarb'.

- Declined to reintroduce Covid lockdown restrictions in December 2020 – determined to avoid the moniker 'The Man Who Cancelled Christmas' – finally doing so in January 2021, after most schools had been back for a single day. The Resolution Foundation estimated that this delay caused 27,000 extra deaths.

Death of supporting characters? All present and correct. The *good old man* is Kenneth Clarke – the longest-serving Member of Parliament and prominent supporter of the EU, forced out in 2019. The *rival and shadow* are David Cameron and Theresa May. The *innocent(ish) children* are the twenty-one MPs Boris sacked for refusing to back his Brexit plans. *The temptress?* Well, I don't want to get sued, but Boris's affairs are legendary, to the point that he refused for many years to confirm how many children he had.[7]

No wonder our Icarus thought he was untouchable. Inevitably, he turned out to have been flying too close to the sun for too long. Surely just another couple of minutes couldn't hurt? After all, compared to everything else that had happened, appointing and backing a colleague with a history of groping accusations (named – you couldn't make this up – Chris Pincher) feels pretty insignificant (compared with, say, being second only to the USA in Covid deaths per capita). Yet Pincher turned out to be the proverbial straw that broke the camel's back, heralding the **destruction**. Boris could easily have sacked Pincher, but for his own fatal character flaw: his need to be liked. The two were mates, and – almost sweetly – Boris didn't want Chris to stop liking him. The die was cast. One hot day in July 2022, a slow drip of government resignations over the affair turned into a flood of melting wax, and the wings gripped nothing but empty air. Boris never lost a confidence vote, much less an election, but such are the unwritten and unknowable rules of British politics that even Boris couldn't fight its gravity, his legs thrashing in impotent rage as he fell through the sky. Alex's character flaw had undone him, as he always knew it would: where do you go if you need to be loved, but nobody loves you? Into the sea.

THE SCIENCE BEHIND THE STORY

Why are Icarus stories popular? While we might enjoy a little Schadenfreude at the expense of Boris Johnson, no normal person takes enjoyment in others' real-world personal tragedies. Even fictional Icarus stories like *Parasite* and *Macbeth* make for grim viewing. Yet they remain not just hugely popular, but critically acclaimed.

Why? Well, as we've already seen, for the 'sweet' variety of the Icarus recipe, it's simple: we enjoy the Schadenfreude (literally, 'harm-joy'), one of those foreign words that captures a notion so perfectly that English hasn't even bothered trying to come up with its own (in Japanese, they say, 'The misfortune of others tastes like honey'). As for a real-world example, Tiffany Watt-Smith, who literally wrote the book on *Schadenfreude*, cites a particularly delicious case: a Louisiana preacher, Tony Perkins, had long been kicking up a storm with his views that natural disasters are God's punishment for abortion and gay marriage. You can guess what happened next. Floods hit Perkins's home, and he had to flee with his family in a canoe. With no apparent hint of irony, our Tony described the flood as being of 'near-biblical proportions'. Schadenfreude researchers (who find it very difficult to get funding, as they just love rejecting each other's grant proposals) have identified three personality characteristics that make an individual ripe for Schadenfreude: *unlikeability*, *deservedness* and *envy*; and Perkins has each of these in spades.[8]

Incidentally, thanks to modern neuroscience techniques, researchers are now able to track Schadenfreude in the brain in real time. In a study conducted at the Institute of Cognitive Neuroscience in London,[9] participants played a gambling game with a partner. As usual in these types of studies, the 'partner' was actually in cahoots with the researchers and had been instructed to either play fair – which resulted in the participant winning money – or to cheat, effectively stealing money from the participant. Afterwards, the partner – not the participant – was given a painful electric shock to the hand.* All of this took place while the participant was in an fMRI scanner, to allow the researchers to monitor the activation of different parts of

---

* A real shock, thank you very much, none of that Milgram nonsense (from the Monster chapter) here!

the brain as the experiment unfolded. What they found was that when a 'cheating' partner received an electric shock, the 'reward circuit' of the participant's brain (specifically, the left ventral striatum/nucleus accumbens) lit up in apparent delight, though only for male participants. In contrast, when a 'fair' partner received an electric shock, the 'pain circuit' of the participant's brain lit up in apparent empathy, though to a greater extent in female than male participants. Indeed, female participants showed some neural evidence of empathy (or 'pity') even when a 'cheating' partner received a painful shock.*

So, yes, it's easy to understand why we enjoy sweet Icarus stories in which an anti-hero gets their comeuppance. But why on earth do we enjoy the more complex flavours of *Parasite* or *Crime and Punishment*, let alone the downright bitter flavours of *Romeo and Juliet*, *Titanic*, *Gone with the Wind* and *Brokeback Mountain*? The question of why we take, in some sense, 'pleasure' from relentlessly sad, painful works of art – movies and novels, yes, but also music, poems, paintings, photos – has troubled just about every famous philosopher, but none of their answers quite hit the mark.

Aristotle himself speculated that fictionalized tragedies may teach us how to behave in similar circumstances. Yes, they probably do, but why should this learning be *pleasurable* rather than painful? David Hume thought that the heightened emotions of tragedies capture our interest and stave off boredom, but that is equally true of both 'sweet' Icarus stories and those following all of the other masterplots in this book. Why do we love a weepie? A lesser-known philosopher, John Morreall, suggested

---

* Of course, this study cannot speak to the question of whether the apparent gender differences (which we must consider tentative, since this is only a single study) are genetic in origin, or a product of the gender stereotypes that participants will have absorbed from early childhood.

that we enjoy the feeling of control we get from knowing we can put down the tragic Icarus story at any point. But then why do we pick it up in the first place? This would be like banging your head against a wall because it feels good when you stop. Many have pointed to our species' thirst for gossip, which – as we saw in Chapter 1 – has no doubt helped us throughout our history to avoid countless dangerous situations and shady characters. But if that's our motivation, the gory details are pure overkill. Who, exactly, is watching *Titanic* 'for information'?

No, these explanations aren't satisfying; mainly – I think – because they seem to somehow trivialize and cheapen the deep emotions we feel when we experience 'sad' works of art. We are *savouring* the sadness for its own bittersweet flavour, not merely *enduring* it because we're picking up some useful information or looking forward to switching off the TV set. More promising, then, are approaches which acknowledge that hedonistic pleasure isn't the only emotion worth having. Most lottery winners don't blow their wad on drink, drugs and prostitutes, and those who try almost always find that the novelty wears off pretty quickly.

But if it's not pleasure, then what emotion are we basking in when we 'enjoy' a full-on weepie? An important clue comes from studies of sad music. When people are feeling sad, what do they listen to in order to cheer themselves up? Upbeat dance music? No, more often than not, sad music. Paradoxically, when you feel sad, sad songs make you feel not sadder, but happier. In a 2020 journal article,[10] the musicologists David Huron and Jonna Vuoskoski proposed that sad music evokes *compassion*, which itself is a pleasurable emotion. Indeed, philosophers including Thomas Hobbes and Immanuel Kant have argued that acts of charity – even anonymous ones – are not completely altruistic, because the giver gets to bask in the pleasurable emotion of 'compassion'. Perhaps, then, we enjoy listening to sad music because it evokes compassion.

It's an interesting theory, but how can we test it? The key, argue Huron and Vuoskoski, lies with the fact that not everybody enjoys sad music, and this seems to be true the world over. These researchers surveyed taxi drivers (some of the world's most strongly opinionated consumers of music) in China, Egypt, Ethiopia, Greece, Haiti, Kashmir, Mali, Nigeria, Pakistan, Punjab, Russia, Serbia, Somalia, Tunisia and Vietnam. Just over half (58 per cent) said that they liked listening to sad music, with the remainder saying they didn't. So, what differentiates those who do and don't enjoy sad music? After running a battery of surveys and personality tests, Huron and Vuoskoski had their answer, and it was exactly as they predicted: people who enjoy listening to sad music score highly on measures of compassion. Those who tend to feel lower levels of compassion don't enjoy listening to sad music, and often find it puzzling that others do. Sad music is enjoyable, then, because it evokes feelings of compassion, and those feelings are themselves pleasurable.

But this is music. What does it tell us about stories? Quite a lot, actually. Listeners who enjoy sad music scored higher not just on compassion, but also on a measure of *fantasy*. That is, they tend to feel absorbed and transported into the narratives of novels and movies. In all likelihood, then, the picture is similar for music and narratives. Those of us who enjoy stories that follow the tragic version of the Icarus masterplot – and, as with sad music, not all of us do – enjoy the feeling of experiencing compassion for the protagonist. As far as I know, nobody has yet directly tested this enjoyable-compassion theory with narratives, but it is very promising in one respect: it offers a plausible explanation of why we find fictional tragedies enjoyable, but real-life tragedies horrific. When we watch a real-life tragedy – say a documentary about starving children – we find ourselves in an unpleasant double-bind. We must either decide to turn away, and so endure feelings of guilt, or to help, which is not only costly, but often causes even more guilt: say we decide to donate £50, couldn't we afford £100? And if not, why not? Do we really need that £2,000 holiday more than these starving children need *food*? Shouldn't we cancel the holiday and donate the money instead? In contrast, fictional tragedies like *Romeo and Juliet*, or stories based on real but historical tragedies like *Titanic*, allow us to enjoy feelings of compassion without the counterweight of guilt: no matter how much we might want to, there is simply nothing we can do to help Juliet or Rose.

UNDER THE INFLUENCE

The Icarus masterplot recipe is unique amongst those in this book, in that nobody would ever choose to follow it consciously. The closest we see perhaps is when people or organizations – while ostensibly aware of their Achilles heel – see this character flaw as a strength rather than a weakness, and double down when they should instead cut their losses. This is seen most easily not with individuals, but corporations.

For all my adult life, I've been in bands. First was the deeply unpopular Starla, which morphed into the marginally less unsuccessful post-rock band Advances in Mathematics, when I belatedly decided to spare audiences – such as they were – my attempts at singing. After stints with Below The Stars (Sixties pop) and The Angel Hurricane (Stone Roses-wannabe indie) came Help Stamp Out Loneliness (indie pop), which did manage to carve out something of an audience, albeit in the corner of the nicheiest niche imaginable. These days, I content myself with occasional cover band gigs. In the background to all of this – or actually, I suppose, the foreground – was a Premier League of more popular, successful and – damn them – talented bands doing the same Manchester indie circuit: Polytechnic, The Longcut, The Answering Machine, even The Courteeners who, to their undisguised dismay, somehow found themselves playing my night at the woefully named Bar XS (now a Sainsbury's cafe).

Whether big (eventually) or small, all of these mid-2000s indie bands had something in common. 'Yeah, we've got a MySpace,' they'd mumble embarrassedly between songs, 'it's MySpace dot com, forward slash – all one word . . .' It's hard to imagine now, but MySpace really was the biggest website in the world for a while.

If you're a Millennial, you almost certainly had a MySpace profile. But if you're any younger, you've probably never heard of it. So just what was MySpace exactly? Basically, it was a social network, quite similar to Facebook (which I know you don't use either, but hey), but it had a Spotify-like element too, in that a high proportion of profiles were bands, and you could listen to their songs on their profile pages. Individuals could also borrow these songs for their own profile pages and even set them to play automatically whenever anyone visited (this would be a big no-no these days, but didn't raise an eyebrow in the mid-2000s). It also had a dating-site element, with teens posting their grainy, poorly lit thirst traps, taken on flip-phone Motorolas and Nokia bricks. Yet MySpace was much bigger than that makes it sound. These days we have social media apps, music apps, video apps, messaging apps, dating apps, and several competitors to choose from in each category. At the peak of its powers, MySpace was Facebook, X, TikTok, Spotify, YouTube, WhatsApp and Tinder, all rolled into one.

THE STORIES OF YOUR LIFE

If you lived through the demise of MySpace, you'll probably be surprised to learn that although it's deep, deep into its **destruction**, it isn't dead, exactly; more undead. If you try to visit your old MySpace page you'll find it's still there, but it's a spooky hollowed-out shell. Your frozen-in-time biography is up, along with flyers for gigs that you went to in 2007, but most of the photos – and just about all of the songs – have been lost over the years in various server migrations and clean-ups. You can click on them, but they won't play. Where did it all go wrong?

The key to understanding any Icarus story is the central character flaw of the main protagonist. MySpace's flaw is that, unlike Facebook, it didn't start out as a social networking site. As Sean Percival, the company's former vice president of online marketing tells it, MySpace actually started out as the project of a company that sold tat online: diet pills, remote-controlled toy helicopters, nothing fancy. Looking for ways to expand the reach of their adverts, they hit upon the idea of copying Friendster, one of the very first social networks. Right from the very start, then, the whole raison d'être of MySpace was to create a network of customers to show adverts to.

Of course, this wasn't the face it showed to users, for whom MySpace was a network of friends. The concept is so everyday now, it seems bizarre even trying to explain it, but it was on MySpace that me and millions of my generation first came across the idea of a virtual friend: someone who is your friend on social media, but who you've never met in person. I made my first friend, Tom, as soon as I set up my MySpace account. Of course, I soon realized that Tom – Thomas Anderson, the founder of MySpace – was set to appear by non-negotiable default as the first friend of all new users. Soon, however, I was sending friend requests to people I knew – or at least half knew – in real life, and much ink was spilled in the serious newspapers about this disastrous trend amongst young people of swapping 'real' for 'virtual' friends, and where would it all end?

The pivotal **dilemma** came in 2005 – just two years after the site's founding – when Rupert Murdoch put in an offer of $580 million (that's almost $3 per friend). In an Underdog story, Tom and friends would have thumbed their noses at the old man. But this being an Icarus story, the sale to Murdoch was inevitable, preying as it did on the character flaw or *Achilles heel* of the protagonist: in this case, MySpace's origins as a marketing ploy. The company's better-than-anyone-could-have-expected **elation** stage peaked in 2006, when it racked up more visits in the US than any other website, including Google.

At this point, **insatiability** reared its inevitable head. Nothing is ever enough; with such a huge investment, the site quickly came under pressure to make money. As the anti-hero always does in this situation, it doubled down, doing the thing that had won it early success, only more so; *much* more so. For MySpace, that thing was showing users adverts. Lots of adverts.

If your only experience of social media is the sites that came after – and learned lessons from – MySpace, you won't believe the adverts. Modern social media sites favour promoted or sponsored content that, often, hardly looks like adverts at all. MySpace was the Wild West, with adverts all over the page and no restrictions on formats. The nadir was an ad called 'Punch the Monkey'; you had to click on the monkey to punch him, then complete a survey, sign up for a credit card and so on to – maybe – win a prize. If you've never seen it, be grateful – it was almost as if someone had been given the task of coming up with an advert that would annoy users as much as possible. This 'anything goes' approach extended to users' profile pages: users could embed music, animated GIFs, website links – anything at all – just about anywhere in their profile page. Sites like pimp-my-profile.com made it their business to ensure that every MySpace page was an unreadable mess of flashing, scrolling text on top of pictures, over a backdrop of music and sound effects. And – again – adverts: no restrictions were placed on what could be left in the 'comments' section of users' profile pages, meaning that they were swamped with adverts and spam.

The seeds of MySpace's **destruction** were sown. Users fled to Facebook in their millions, and Murdoch was eventually forced to sell for a reported $35 million.

In 2021, researchers from the University of Cambridge's business school published an investigation of the companies that were listed on the London Stock exchange in 1948.[11] How many, they asked, manged to survive until 2018? Half? A quarter? A tenth? The answer is just over 1 per cent: 19 out of 1,513 firms. To a first approximation, then, the tragic tale of MySpace is the fate of all companies.

What keeps going wrong? The popular view is that long-established companies are rabbits in the headlights that are paralysed into inaction when a threat arises. But writing in the *Harvard Business Review*, the economist Donald Sull argues that this isn't quite right.[12] Big companies know their markets well: they typically pick up on threats early and act. It's just that they almost always do the wrong things. Sull calls this 'active inertia'. This isn't failing to move at all (that's 'passive inertia'), but doubling down on all the things that have worked well in the past.

Does this sound familiar? It should. Most corporate stories end in tragedy because when the dilemma comes, and firms must choose between the 'right' path and the 'wrong' path that appeals to the Achilles heel, the character flaw, the ethos of the company, they almost inevitably – like the anti-hero of an Icarus story – choose the latter. Sull discusses the example of Firestone Tires, whose defining ethos was 'family values'. For example, the founder Harvey Firestone proudly and magnanimously set up the Firestone Country Club, open to all members of the company from the CEO to the humble factory worker. Local plant managers were fiercely loyal to their teams, and were encouraged to bid for additional capacity at their plants, with their requests usually waved through by chummy executives. Firestone was a family firm in the most literal sense: at the point when the crunch came, in 1972, *one third* of the executive board were the children of former board members.

The crunch came in the form of Ford, Firestone's largest customer, mandating that all its cars would from now on use radial tyres, a safer and longer-lasting type recently developed by the French company Michelin. Firestone knew this development was on the cards, and did invest in new capacity for making radial tyres, though mainly by tweaking its existing operations. This was nowhere near enough. What it desperately needed to do was to bring in fresh expertise, and close down plants that were still making the old tyres. But its corporate family values – however laudable in the abstract – more or less prevented it from doing so. The path that appealed to the ethos of the firm was always the one that it was going to take, but it was also the one that led to ruin.

A similar fate befell the British clothing and home furnishings firm, Laura Ashley, which was characterized by a rejection of the sexualization of women's fashion and an appeal to traditional British values such as modesty and restraint, epitomized by a kind of rural idyll. These traditional values were key to the company's success, but became a millstone around its neck as the world changed and more and more women entered the workplace. Shoulder pads and power dressing were in. Rural idyll was definitely out.

If you run a business, whether it's a multinational corporation or your own neighbourhood coffee shop, you are no doubt wondering how to avoid falling prey to a MySpace-style corporate tragedy. As we've already seen, it's not easy – most firms fail sooner or later – but we can learn some lessons by following the example of . . . McDonald's.

Yes, surprising as it may seem, the global megabrand was in serious trouble in the early 1990s, and the cause was a familiar one: relentless doubling down on its core ethos, its personality. Or perhaps, in this case, its lack of personality, since McDonald's' ethos is consistency and predictability, across both time and space. The chain prides itself on the fact that the Big Mac that today tickles your tastebuds in Tokyo is identical to the Big Mac that you munched on in Moscow at the turn of the millennium (it is – I've tried both!).

Like MySpace's materialism, Firestone's family values and Laura Ashley's aesthetic, McDonalds' core characteristic – uniformity – became a liability; its Achilles heel. Customers were bored of burgers – they wanted new, healthier options. McDonald's tried to innovate, but were so used to thinking inside the burger box that their new offerings – the low-fat McLean burger, and the adult-oriented Arch Deluxe burger – were widely seen as more of the same; just a bit worse.

Beginning at the turn of the millennium, though, McDonald's turned it around. How? They realized that they had to change their core value – radical homogenization – before it led to death and destruction. Nowadays, McDonald's sells poutine in Canada, mango McFlurries and Thai green curry chicken burgers in Malaysia, dosa masala and spicy paneer burgers in India, shrimp burgers in Hong Kong, red bean pie in China, bulgogi burgers in South Korea, Big Brekkie burgers in Australia, the halal McArabia Chicken in Saudi Arabia, spaghetti in – God, no, not Italy – the Philippines, and teriyaki samurai pork burgers in Thailand, not to mention salads, wraps, McCafé . . . the list goes on. McDonald's have shown that following the Icarus masterplot recipe doesn't have to be inevitable. If you have enough self-awareness to identify and work on your Achilles heel, you can change the narrative and escape tragedy.

Would that work for people too?

PLOT TWISTED

> The race is not to the swift or the battle to the strong . . . but time and chance happen to them all . . . No man knows when his hour will come: As fish are caught in a cruel net, or birds are taken in a snare, so men are trapped by evil times that fall unexpectedly upon them. *Ecclesiastes, 9:11–12.*

Generally, when a masterplot is abused, misused or perverted, this is done deliberately for nefarious ends: whether that's winning elections, convincing you the moon landings never happened or simply selling you more tat. The Icarus masterplot is an exception in that – invariably – the perversion of this plot occurs when someone who has been through tragic real-world events just can't help seeing them as conforming to the Icarus recipe when, objectively speaking, this is not the case.

It's easy to see why this happens. After all, a major theme of this book is that we humans attach one of a small number of masterplots to essentially all of the experiences that we go through, to the extent that we allow them to influence our decisions and emotions. And it's easy to understand why somebody who has been through a particularly tragic life event – say the death of a partner or family member – is particularly motivated to find some kind of meaning, some kind of narrative that helps them make sense of it all.

Take, for example, the death of an unborn child. We can see

straight away that even our choice of terms encourages one or other narrative framing. You probably did a bit of a double take at 'death of an unborn child', wondering why I didn't use the more common term 'miscarriage'. The reason is that 'miscarriage' is a loaded term, suggesting that the person who has 'miscarried' is in some way responsible; that they have 'carried' the unborn child incorrectly (the meaning of the prefix 'mis-', as in *mishandled*, *misunderstood*, etc.). The preferred term these days is 'natural pregnancy loss', designed to emphasize the fact that this is what should, naturally, happen for an embryo that cannot survive out of the womb.

Natural pregnancy loss is more common than most people realize, affecting perhaps around 30 per cent of pregnancies (counting across both known and unknown pregnancies).[13] But we're focusing here on pregnancy loss that is not diagnosed until the first scan, at ten to twelve weeks (around 1–2 per cent),[14] which is an extremely traumatic event. In fact, a 2005 study conducted in London found that associated levels of grief were as high as for those who had been through the death of a close relative.

As we have seen throughout this chapter, stories that follow the Icarus masterplot always end in death and destruction; but not just any old death and destruction – death and destruction that could have easily been avoided if only the protagonist hadn't pushed things too far; hadn't – against their better judgement – chosen the path of temptation when presented with a dilemma that targeted their Achilles heel.

The real world, of course, is not like that; or at least not always. Bad things happen to good people, through no fault of their own, while the guilty get off scot-free. The problem is that even though – on some level – most of us know this, we struggle to accept it. So enthralled are we with masterplots, that we can't help interpreting real-life events through their recipes.

Icarus, sadly, is no exception. As a result, almost all those who have experienced pregnancy loss experience at least some level of self-blame. 'I shouldn't have gone on that holiday' they say, identifying (incorrectly, of course) the temptation to which they succumbed. 'I wasn't sure about it at the time' (dilemma), 'In fact, I knew it was the wrong thing to do' (transgression), 'but I look forward to my summer holidays all year' (Achilles heel). 'I had a good time' (elation), 'but I obviously overdid it' (insatiability), and 'now I'm paying the price' (destruction). Of course, it's not just pregnancy loss. People who have, to take just one more example, lost a parent often go through a similar internal monologue in which they blame themselves for sending – or not sending – the parent to hospital or a care home, or for not taking their concerns seriously, or for not visiting them enough . . . Worst of all, true to the Icarus masterplot recipe, some will see their actions as reflecting some deep-seated character flaw – 'I'm too reckless/selfish/inconsiderate/tight with money, etc.'

We know this in the folk wisdom that we try to pass on to the bereaved: 'You mustn't blame yourself,' we say. 'You'll drive yourself mad.' This advice is more on the mark than most of us probably know. 'Madness', of course, is not a term used in modern medicine (indeed, many would consider it offensive). Instead, psychologists and psychiatrists talk about 'psychosis' – disordered patterns of thinking and perception that, at their worst, involve losing touch with reality – a common symptom of schizophrenia and related severe mental health conditions. Two of the main symptoms of psychosis are delusions and hallucinations, which were long considered symptoms of – as it was called in less enlightened times – 'madness'. These symptoms are similar but distinct: delusions are thoughts that aren't true; hallucinations are sights, sounds or even smells that aren't there. For example, perhaps the most famously delusional character in literature is Shakespeare's King Lear, who at one point asks, 'Am I in France?' ('In your own kingdom, sir,' the Duke of Kent replies). Perhaps the most famous literary hallucination is the blood on Lady Macbeth's hands ('Out, damned spot'), or her husband's vision of a dagger before he kills Banquo.

In 2019, a group of psychologists based in Hamburg, Germany published a systematic review and meta-analysis drawing together the results from all available studies that looked at the relationship between self-blame and these psychotic symptoms.[15] What they found was that the more self-blame patients showed (as measured by questionnaires), the more often they experienced delusions and hallucinations. Moreover, patients with psychosis showed significantly more self-blame than healthy controls. Of course, these findings are only correlational – we don't know for sure whether the self-blame *caused* psychosis or vice versa. But they are, at the very least, *compatible* with the old folk wisdom that blaming yourself will – in the common vernacular – 'drive you mad'.

Psychosis is an extreme example, but there is evidence from a wide variety of studies that self-blame – applying the Icarus narrative whereby you yourself caused the tragedy – makes everything worse. As any parent will tell you, there can be no greater tragedy than the death of a child. Sadly, but unsurprisingly, another systematic review found that around half of bereaved parents do blame themselves, and that this self-blame results in worse outcomes for the parents.[16] Parents who blamed themselves for the death of their child were almost twice as likely to show persisting symptoms of anxiety, and almost three times as likely to experience clinical depression. Similar findings were observed in a longitudinal study of teenage sexual abuse survivors: the more self-blame the victims reported when they began their treatment, the greater the clinical depression they showed six months later.[17] Importantly, the longitudinal nature of this study – i.e. the victims completed the self-blame questionnaire at Time 1, and the depression questionnaire later, at Time 2 – means we can be fairly certain that greater self-blame *caused* greater depression, rather than vice versa.

This is bleak, bleak stuff; but could there be just a chink of light?

HAPPY ENDINGS

So far, when it comes to real-life tragedies – the sympathetic kind, that is, not the Boris Johnson Schadenfreude kind – we have seen the Icarus masterplot as a thoroughly bad thing; something that causes innocent people to blame themselves. But there is a more positive side. Once we recognize that someone is misapplying the Icarus masterplot to their own life circumstances, we are halfway towards helping them.

Encouragingly, a few studies have started to investigate whether it is possible to change the narrative of self-blame and, in so doing, lead to more positive outcomes. Self-blame is common amongst medical students who must take life-and-death decisions while learning on the job and who, almost inevitably, make at least some mistakes (hopefully only minor ones). Researchers in Malaysia developed an intervention centred around – corny though it sounds – treating stressful events as learning opportunities.[18] Compared with a control group, medical students who received the intervention showed lower levels of self-blame and – apparently as a result – lower levels of depression and stress.

That's promising, but it's fairly low-level stuff. Junior doctors, after all, know what they're getting themselves into, and that some mistakes are par for the course. What about the rest of us? Well, a study conducted in Australia evaluated an intervention for mothers who had gone through a particularly difficult experience of giving birth, placing them at risk of psychological trauma.[19] The intervention was, at face value, simple – a counselling session with a midwife within the first three days of the birth and again a month or so later. (You would be forgiven for assuming that such counselling for traumatic births would be routine; but apparently not, at least in many parts of the world.) An important aspect of this counselling involved discussing the mothers' views on whether anything should have been done differently during labour, and 'gently challenging and countering distorted thinking such as self-blame'. The results were encouraging: compared with a control group, mothers who received the counselling intervention showed lower levels of self-blame three months later and – apparently as a result – fewer symptoms of psychological trauma and greater confidence around future pregnancy.

Most promisingly of all, remember that study which reported that levels of grief after natural pregnancy loss were as high as for those who had been through the death of a close relative?[20] Well, the researchers in this study didn't just measure grief and self-blame; they set out to do something about it. What they did was fairly simple: just telling mothers exactly what had caused the pregnancy loss,[21] and comparing their outcomes with mothers for whom the cause could not be precisely identified.* You will be unsurprised to learn that when the researchers did

---

* This isn't a perfect method, as there may have been other differences between the two groups. That said, the researchers did confirm that there was no obvious difference between the two groups, such as – for example – the women whose pregnancy loss lacked an identifiable cause being older. Neither did they have noticeably different pregnancies.

manage to identify the reason for the pregnancy loss, it was never 'overdoing things a bit', stress, going on holiday, having sex or any of the other things that people often blame themselves for. Rather, every loss was due to either chromosomal abnormalities or immunological issues.

Crucially, this information alone was healing: at a four-month follow-up, 20 per cent of mothers whose pregnancy loss could not be explained were still in the highest-level grief category, as opposed to just 9 per cent of mothers who had been given an explanation. Importantly, it's the explanation itself that seemed to help, with mothers in the latter category showing significantly lower levels of self-blame on a questionnaire. Simply hearing the message 'No, it was not some temptation that you gave in to, it was . . .' helped these grieving parents to reject the Icarus narrative of self-blame.

While the more extreme examples we've met in this chapter are things that most of us will mercifully never have to experience, a sobering thought is that if you are in a committed long-term relationship, one of you will almost certainly live through the death of the other ('Till death do us part', as they say). Again, key to your – or your partner's – recovery will be the extent to which you are able to step outside the self-blame Icarus narrative that you have internalized via years of reading, listening and viewing.

Unsurprisingly, given the studies that we met above, evidence from a recent study – this one conducted in the San Francisco Bay Area – found that the more the bereaved blame themselves, the more they experience symptoms such as failing to accept the death or continuing to cry every time they think of their departed partner, even five years later.[22] But there's a twist: some of the quickest recoveries were shown not by those who rejected the Icarus narrative altogether, but who spun it on their dead partners ('Well, the doctor did tell him to stop drinking!'). Macabre though this sounds, blaming the deceased partner for their own death allowed the bereaved to make sense of the loss via a meaningful narrative – here, an Icarus story – while absolving themselves of any lingering blame.

Such, then, is the power of masterplots, that even this – on the face of it – most depressing of recipes can allow us not only to find meaning in our evoked feelings of compassion, but even to face head-on, in the death of a loved one, just about the most traumatic experience that any of us will ever have to face.

5.

MONSTER

NOW the forces of destruction gathered together their armies to battle.

And there went out a champion of the camp whose height was six cubits and a span.

And all the men of the city, when they saw the man, fled from him, and were sore afraid.

And the men said 'Have ye seen this man that is come up? The man who killeth him, the mayor will enrich him with great riches, and give him whichever daughter of the town that he so desire.'

Now there lived in this town a young man who had been cast out by his employer owing to his wanton idleness and idolatrous ways.

And this man came up to the rooftops and shouted for battle.

And when the giant looked about and saw the man, he disdained him, for he was but a youth, and ruddy, and of a fair countenance.

And the youth put his hand in his bag, and thence took a magical weapon devised by his brethren, and smote the giant; and the giant fell upon his face to the earth.

And the men of the town arose and shouted, and there was unto them joy and rejoicing, and they drank wine with a merry heart.

This ancient and historic text is, of course, the finale of the 1984 movie *Ghostbusters*. But so faithfully does it follow the template of the (overcoming the) Monster masterplot recipe that I was able to take it, almost word for word, from the Old Testament story of David versus Goliath (OK, so I had to change the odd word here and there, but the underlying story is very much the same).

Ghostbusters was far and away my favourite childhood film. Sadly, I'll never be a Ghostbuster. But, hey, at least I share their original day jobs. That's right – Peter, Ray and Egon start the movie as psychology professors, in their case at Columbia University in New York. Our heroes' call to action comes, quite literally, in the form of a phone call from an administrator at the New York Public Library, where a ghost has been terrifying an elderly librarian named Alice, by flinging around books and index cards. They take a few readings, and even a slime sample, before heading back to the office, only to find they've been dismissed from the university for their slapdash approach to science.

There's only one thing to do – take out a third mortgage on Ray's childhood house and put the money into their start-up, Ghostbusters. Having shelled out on TV adverts, the Ectomobile and a hip-if-dilapidated downtown HQ, the Ghostbusters are down to their last dollar, and are saved from financial ruin only when they earn their first pay check: $5,000 from the Sedgewick Hotel for capturing a ghost affectionately known as Slimer.

Lurking in the background to all these high jinks is a shadowy, shapeshifting god of destruction known as Gozer, who gradually begins to make her presence felt. Peter visits a former Ghostbusters' customer, Dana, for what he thinks is a date – and he's almost seduced by her, but gets the ick when he realizes she's been possessed by one of Gozer's ghostly servants, Zuul. Meanwhile,

Walter Peck – an inspector from the Environmental Protection Agency – has long had (frankly, quite understandable) concerns about the Ghostbusters' homemade containment unit (where they keep all the ghosts they've trapped) and orders it shut down. Escaped ghosts swarm New York – driving taxis, eating hot dogs and generally living their best lives over a pumping 1980s synth soundtrack.

Thanks to Peck, the Ghostbusters are in jail, but fortunately managed to smuggle in the blueprints of Dana's swanky apartment building – 550 Central Park West – which, they now figure out as bemused cons look on, is some kind of portal through which Gozer will enter our world in the form of a destructor. Yikes! Much to Peck's disdain, the mayor allows the Ghostbusters out of jail for one last showdown. Gozer appears in the form of a giant marshmallow man after inviting our heroes to 'choose the form of the destructor' (Ray can't help but think of something apparently harmless from his childhood: Mr Stay Puft, a character who appears on the packaging of his favourite brand of marshmallows).

Victory seems certain for Gozer, as Mr Stay Puft crushes cars and buildings beneath his mallowy feet. Egon, who had previously told the Ghostbusters never to cross the streams of their proton packs, suggests that – given the circumstances – it might be worth a try after all. The resulting total protonic reversal destroys Gozer's portal to the underworld, and with it the marshmallow man and much of the Upper West Side. The hero gets the girl – a now de-Zuuled Dana – and the whole city comes out to wave and cheer. Cue the music.*

---

* And not just any old music: for my money, the best movie theme ever (and, yes, that includes 'Live and Let Die', Mr McCartney).

THE STORIES OF YOUR LIFE

THE MASTERPLOT RECIPE

Six key ingredients give the Monster flavour to what is, underlyingly, a very basic recipe that follows the three-act structure we met in Chapter 1 (*Inside the Story*). The hero is at home minding their own business (conducting paranormal psychology research at Columbia University), something happens that upends the normal order of things (Gozer opens up a portal to the underworld), then, finally, this unstable situation is resolved (the Ghostbusters defeat Gozer, sealing the portal).

The most central of these key ingredients is the **monstrous monster**. What I mean by this rather tautological-sounding description is that the monster can take almost any form, except that of a run-of-the-mill human. It must be, in the literal sense of the term, 'supernatural' – above or outside of nature. Often it will take a humanoid form, but it must deviate in some way. A humanoid monster must be abnormally large (Goliath, the Stay-Puft Marshmallow Man, the giant in Jack and the Beanstalk, the ogres of fairy tales) or small (dwarves, goblins, elves etc); it must have supernatural powers (Count Dracula, Sauron, your bog-standard witch); it may be a creepy human-animal hybrid – the Minotaur, harpies, Medusa. If you were really out of ideas, you could – in less enlightened times – simply give your human baddie some kind of disability or disfigurement. In fact, the James Bond franchise is still stuck in the 1970s in this respect, with both of the villains in *No Time to Die* (2021) having facial scars (to its credit, the British Film Institute has launched a campaign against what it called this 'lazy stereotype').[1]

If you don't fancy a human(oid) monster, you can have an animal; but again, it can't just be your bog-standard cat or dog, it must be a giant, a chimera (the classic Greek version was a lion with a snake for a tail and a goat's head sticking out of its back), a transforming creature of the night (vampire bat, werewolf) or a fire-breathing dragon. If even that isn't exotic enough for you, the monster can be an alien as in – um – *Alien* or *Invasion of the Body Snatchers*; a plant, as in *The Day of the Triffids*; or even just some mysterious presence that we never see at all, as in the 2018 movie *Bird Box*. These examples capture a final characteristic of the monstrous monster: it can't be just a straight-up baddie who fights fair in an 'honour amongst thieves' kind of way. It has to be not only slippery, devious, cunning and treacherous, but mysterious and hard to pin down in a physical sense: it changes its shape or substance, it transforms from one creature into another, it disguises itself as something innocent, like a marshmallow man, or even one of the goodies.

Yin to the monstrous monster's yang is the key ingredient of the **unlikely hero**. If you know your Bible stories, you'll recall that David was a humble shepherd, and there was widespread disbelief when he was chosen over his older and stronger brothers to fight Goliath. The Ghostbusters were psychology professors, and as someone who knows quite a few, I can confirm that they are literally the last people on earth you would choose to take on a giant marshmallow man.

Like oil and water, the ingredients of the monstrous monster and the unlikely hero don't mix well. It's not a fair fight: surely this all-conquering giant will simply swallow the humble shepherd/farmhand/psychology professor with a single bite? The solution comes in the form of the third key ingredient: the **magic weapon**. The Ghostbusters, of course, have their proton packs, which prove key in their final showdown with Gozer; David has his sling; the *Game of Thrones* heroes have their weapons of Valyrian steel or dragonglass. Usually, before the main action gets going, our hero(es) will pay a visit to some kind of semi-magical armourer who will dispense this magic weapon. This trope is leaned on particularly heavily in the James Bond series, in which Bond always visits the mysterious Q to be given some fancy gadget such as a cigarette gun, a camel saddle with a blade sticking out of the seat or a hovering tea tray that can slice off the enemy's head (all real examples!). *Ghostbusters* follows this tradition by ensuring that the central character of the trio, Peter (the Bill Murray one) plays no part in the invention of the proton pack, leaving Ray and, in particular, the geeky Egon to play the role of armourer.

Another key ingredient of the Monster masterplot is one that, in homage to my favourite childhood movie, I've named after the *Ghostbusters'* strapline: They're **here to save the world**. In every Monster story, the monster is threatening not just our hero, but – at the very minimum – the whole city; more usually, the whole world, our way of life or all of humanity. Whether it's a Bond villain, Gozer, Darth Vader or Count Dracula, the monster's goal is to conquer the entire world (or, in Vader's case, the entire galaxy). Consequently, the hero is fighting not to save his own skin – he needn't personally have taken on the challenge at all – but on behalf of his city, his tribe, his species or the entire planet.

The Monster masterplot is a flexible recipe, in which the battle between the monster and the hero can play out in a virtually limitless number of ways, with more twists and turns than a particularly challenging Grand Prix circuit. In the end, though, the hero always kills the monster. Sorry for the spoiler, but this ingredient – **slaying the dragon** (or marshmallow man, giant, ogre, dwarf, witch, Bond villain, alien, body snatcher . . .) – is absolutely non-negotiable. If this doesn't happen, it's not a

Monster plot (and if it ends with the monster victorious, it's almost certainly an Icarus).

The final ingredient of the Monster plot is technically optional, but rarely missed out: the **hero's reward**. Unfortunately, since literature – like everything else – was *extremely* sexist until about five minutes ago, this is more often than not, the hand – to put it *extremely* euphemistically – of the sexiest young woman in town (in *Ghostbusters*, Sigourney Weaver's Dana). They probably didn't tell you this part in Sunday school, but, yes, David's reward for killing Goliath was not a new harp, but Princess Michal (daughter of King Saul). And the less said about the James Bond movies the better.

I'm sorry, I don't make the rules, OK? But like a pinch of salt, the reason that most Monster recipes add this ingredient – aside from minor vicarious titillation for the heterosexual males in the audience – is to bring out the flavour of another ingredient: in this case, **here to save the world**. The hero's reward emphasizes that the hero was fighting not solely for his own skin, but for the good of the whole community, who must therefore express their gratitude.

But just why are **here to save the world** and all the other key ingredients of the Monster masterplot required in the first place? We'll start to find some answers to this question in a few pages time, when we delve into the science behind the masterplot. Before we can get to grips with this science, though, we first need to understand how the Monster masterplot plays out in the real world.

This might seem like a strange thing to say. But although none of us, of course, encounter *literal* monsters, most of us come face to face with some or other dark, mysterious, all-powerful, almost supernatural force; that is, with a monster.

STRANGER THAN FICTION

The actor Claudia Christian has appeared in over forty TV series (including *Dallas*, *Babylon 5*, *Columbo* and *Blood of Zeus*), over fifty movies (including *Hexed*, *The Chase* and *Atlantis: The Lost Empire*) and seventeen video games (including one of my favourite ever games, *Fallout 4*). In 2017, she gave a moving and powerful TED talk which set out what she called her 'nearly decade long battle with something I refer to as "The monster"':[2]

> Addiction is a monster, and it affects every ethnicity, social class, race, sex, age; it doesn't matter. You can be the most disciplined person in the world. When it gets you, it has you. 'It' is in control.

Addiction as a 'monster' is only a metaphor of course, but it's one that has been life-saving for Christian and many others; hence the popularity of addiction self-help books such as *The Monster* (Julie Hernandez), *The Addiction Monster doesn't live here anymore* (Mikki Alhart), *Slaying the Addiction Monster* (Sheryl Letzgus McGinnis) and *Starve the Monster* (Hugh Quigley).

The thing is, until the 1960s – which saw the publication of a book called *The Disease Concept of Alcoholism*[3] – alcoholism and other kinds of addiction were seen as a moral failure, with weak-willed addicts deserving of our scorn and punishment, not our pity. The monster metaphor rejects this framing entirely. The addict is not responsible for their own plight in any way. A **monstrous monster** has decided to target them, and the addict is no more responsible for the monster than the citizens of New York were for the marshmallow man. Framing addiction in terms of the Monster masterplot also emphasizes that the power of the monster is, in some way, supernatural. Addicts cannot, as non-addicts often advise them, 'just stop'; they have a powerful sense that they are, as Christian describes it, 'not in the driver's seat . . . not in control'.

Listening to Christian speak, there is a real **here-to-save-the-world** sense that she's fighting a battle not just to cut down on her drinking, or to be able to enjoy the odd glass of wine at Christmas, but for her very existence: 'I started to suffer from seizures in my body. I lost all control of my motor controls. I couldn't stand up; I couldn't get dressed.'

And Christian, as she tells it, is an **unlikely hero** of her own Monster story.

> I wasn't drinking because I had a crummy childhood or because I was suffering from any personal trauma . . . If you look at it from the outside, I had a great life! I was in my chosen career. I had a beautiful home. I had friends and family who loved me and supported me.

This privileged lifestyle did, however, give Christian an advantage over most addicts: she had the money, time and resources to leave no stone unturned when searching for her **magic weapon**:

> cold turkey . . . equine therapy . . . rehab for $30,000 to basically drink wheatgrass and do tai chi . . . talk therapy for over two and a half years for two hundred bucks a session . . . a hypnotherapist who claimed that he had cured a member of the Grateful Dead . . . twelve different meetings of AA [Alcoholics Anonymous] in two different countries . . . I went macrobiotic. I got my chakras realigned . . . I tried veganism . . . I prayed until my knees were black and blue.

None of it worked, and from a Monster perspective, it's easy to see why. None of these were really magic weapons designed to target the monster – Christian's addiction – itself. All of these are general self-improvement treatments used to treat a wide variety of ailments, or no ailment at all. They don't target the monster, but Christian herself, with vague ambitions to give her the inner strength, resolve or general health to walk away from the monster. It's as if, rather than arming the Ghostbusters with their proton packs, David with his slingshot or James Bond with his cigarette gun, we simply put them on a healthier diet, asked them to do a spot of meditation and then sent them off to battle.

But what makes Christian's story different from the countless other addiction stories that you've heard is that she eventually found her magic weapon. And it really did work like magic:

> I poured myself a glass of wine, and it was a miracle . . . the wine just sat there while I ate my dinner . . . There was no head games, no compulsion. I took a couple of sips and I went, 'Meh. I'm done.'

An hour beforehand, Christian had taken naltexrone, an opioid blocker that blocks the pleasurable effects of (amongst other drugs) alcohol. Unlike the tai chi and wheatgrass, this magic weapon really did target the monster – the physical, chemical addiction – rather than simply trying to gee up the hero for the fight. As a result, Christian really did **slay the dragon**. As she puts it, 'The monster is no longer in control. I am.'

Her **hero's reward** was not just the resumption of her career, but the launch of a new one, as an author and advocate for the treatment that saved her life. Much to Christian's frustration, the Monster narrative is still not the dominant one amongst many professionals:

Studies show that tough love and humiliating an addict or making them hit rock bottom is not helping them; it's actually making people worse . . . If we addicts had a 'normal' disease, we would be treated with sympathy and comfort; instead, we're faced with a barrage of 'Why can't you just quit? Just say no?'

And when you think about it, it is kind of remarkable that a treatment that has been around since the mid-1990s is used so rarely. How many lives, then, could be saved if we followed Christian's advice to stop blaming addicts, and to frame addiction in terms of a Monster masterplot recipe, in which our job is to provide addicts with the magical weapons they need to overcome their personal monster?

THE SCIENCE BEHIND THE STORY

As we saw above, the single most important ingredient in the Monster masterplot is that of the **monstrous monster**: the monster must be in some way supernatural – beyond or outside of what is found in nature. In fiction, it must be some fantastical creature. In Claudia Christian's story, the monster of addiction is supernatural in the sense that – unlike, say, lateness or rudeness – it's an aspect of the addict's personality over which she has no control.

But *why* can't a fictional monster be just a 'normal' person? Why can't a metaphorical monster be just a 'normal' problem? To answer this question, we need to delve into the psychology behind the Monster masterplot.

As readers or viewers, we are invited to identify with the **unlikely hero** through whose eyes the story is told. This is why, as we have already seen, the hero himself (yes, unfortunately, it is usually a 'him') must be extraordinary only in his ordinariness. He must be the 'man in the street' with whom we can all identify; and if he is in some way downtrodden, put-upon or overlooked (like the Bible's David or *Ghostbusters*' bumbling professors), so much the better. So hard and fast is this rule that even when the plot requires the hero to also have some kind of supernatural ability – as in *Superman* or *Spider-Man* – he must don his costume only when on his way to a job, and otherwise maintain his everyday life as 'one of us'.

The reason that the monster must be – well – monstrous is that it must be set up as the absolute polar opposite of the hero both psychologically and – to hammer home the point – physiologically. The hero is generous, kind and selfless. The monster is mean, cruel and selfish; but not just in the way that we humans are all, at times, mean, cruel and selfish. The monster is mean, cruel and selfish in a way that is off the scale, dialled up to eleven, in a way that even the worst human could not possibly hope to emulate. The reason why the monster must be beyond all possible human levels of monstrosity, then, is precisely because the hero and the monster are mirror images of one another: the meaner, crueller and more selfish the monster, the more generous, kind and selfless the hero. And by 'the hero', remember, we really mean 'you, the reader or viewer'.

In other words, then, the more monstrous the monster, the greater the extent to which it reinforces our own view of ourselves as thoroughly good people. And – here comes the science bit – in the hierarchy of human need, our need to think of ourselves as good people is right up there with food, water and a 5G data signal. A well-known and very robust finding in psychology is the better-than-average effect, first applied to narratives in Will Storr's book *The Science of Storytelling*.[4] If you ask people to rate themselves on, say, their attractiveness, their driving ability or their intelligence, most will confidently say that they are better than average. Of course, *by definition*, only 50 per cent of people are better than average (or worse than average) on whatever it is you are measuring. The better-than-average effect applies to just about everything you could ask people about, but it is particularly strong for personality. A landmark 1989 study asked college students to rate (a) themselves and (b) 'the average college student' for a total of 154 different personality characteristics; some good (e.g. *cooperative, responsible, respectful, sincere, kind, clean*), some bad (e.g. *belligerent, humourless, mean, maladjusted, incompetent, shallow*). And whaddaya know, on a seven-point scale, students on average rated themselves one point higher than the mythical 'average student' on the good characteristics, and one point lower on the bad ones. (Interestingly, the only traits on which students did tend to rate themselves close to average were those which are neither unambiguously good nor unambiguously bad; e.g. *neat, religious, obedient, radical*.)

The reason we overestimate our good characteristics and underestimate the bad ones seems to be that we are quick to judge others based on outward appearances ('That guy cut me off – he's belligerent, mean, unethical, ill-mannered . . .'), but – because we are privy to our own circumstances and inner feelings – we are just as quick to rationalize away our own transgressions ('OK, I did cut that guy off, but I didn't want to be late picking up my friend, since I'm considerate, responsible, reliable, dependable . . .').[5]

Crucially, though, this tendency isn't just pointless vanity, but a valuable self-defence mechanism. The only group who don't tend to overestimate their good qualities and underestimate the bad ones are people with depression. Maintaining an unrealistically positive view of yourself seems to be important for your mental health. Monster stories do just this, because they invite us to recoil in disgust from the monster, who is everything we are not (*insecure, belligerent, humourless, lazy, vain, gullible, dishonest . . .*), throwing our positive characteristics into sharp relief.

As a result, when we encounter – in the real world – people who have done unspeakable things, we view them as not just evil and inhumane but, in some sense, actually *not human*.

UNDER THE INFLUENCE

As we've seen, the reason that the Monster masterplot is so popular and powerful is that it encourages us to believe – not necessarily accurately – that we, like the hero of the story, are of better moral fibre than the (literally) inhuman, selfish monster against whom our hero is pitted.

What happens, though, when we seek to apply this plot recipe to the real world? Those who work in the criminal justice system – from detectives and solicitors to court staff and rank-and-file police officers – frequently interact with people who have committed horrendous acts. Naturally, our default response in these situations is to treat these offenders as inhuman monsters. But is this framing helpful, or counterproductive?

Professor Laurence Alison is bald with a longish grey beard. Dressed casually in a surfing hoodie, he speaks softly, but in a way that gives him more rather than less authority. He whispers, because he has the authority not to need to shout. Interviewing Alison – perhaps the world's leading expert on interviewing technique – could have been a bit intimidating, but his calm manner instantly puts me at ease. Chatting amiably from his spare room in North Wales, a large blue whale stuffed toy looming over his right shoulder, he comes across as the exact opposite of the stereotypical police interrogator. And that, it soon becomes clear, is exactly the point. 'There's a reason this person has ended up opposite you, and it's not just because they're evil. If you're not interested in what that is, you're not going to be a good interrogator.'[6]

Picture a police interrogation in a movie: the interviewing officer bangs on the desk. He yells at the suspect, threatens him, calls him a 'piece of shit', maybe even punches him (Jack Bauer in *24* is perhaps the best-known and most notorious example). Until surprisingly recently, this wasn't far off the mark. In the UK, laws mandating that all interviews be recorded and that all suspects have the right to have a solicitor present were passed only in 1992, following two high-profile miscarriages of justice. In the US, it was only in 2009 that President Obama signed an executive order banning the use of waterboarding – pouring water on a suspect's face to simulate drowning (as depicted in Robert De Niro's movie *The Good Shepherd*) – and other 'Enhanced Interrogation Techniques'. It's long been known that these kinds of techniques often lead to false confessions.[7] So why did police persist with them for so long (in many places, even to this day)?

The answer, of course, lies with the Monster masterplot which detectives, like the rest of us, have soaked up since childhood. The suspect is an inhuman monster, pure and simple. There's no point in asking him *why* he did what he did; no one could ever offer any kind of sensible explanation or rationale for such heinous crimes, other than pure evil. The detective is a hero, doing God's work, fighting for what is right and good. No wonder a certain type of detective ('the shit ones.' says Alison) loves the 'traditional' approach. But look what happens when you use it. The following are verbatim quotes from suspects, taken from Alison's files of police counterterrorism interviews:[8]

'You don't know how corrupt your own government is – and if you don't care, then a curse upon you.'

'The purpose of the interview is not to go through your little checklist so you can get a pat on the head. If I find you are a jobsworth, we are done talking, so be sincere.'

'Tell me why I should tell you. What is the reason behind you asking me this question?' [Interviewer: 'I am asking you these questions because I need to investigate what has happened and know what your role was in these events.'] 'No, that's your job – not your reason. I'm asking you why it matters to you.'

Confrontation puts up a wall. Alison, who has also written a book on family relationships, offers an example closer to home. If your teenage child breaks a curfew, and you confront them, they'll argue back, until one or both of you ends up storming off. It's just the same with a suspect. If you insist they tell you what they know, they'll keep silent. There has to be a better way, thought Alison.

Or rather, the Alisons: Laurence and his wife Emily are a husband-and-wife team. Emily came from a background in counselling which, the couple soon realized, overlaps with interrogation more than you might at first think. In both cases, the aim of the game is to get the interviewee to open up about something that they'd rather not talk about. The parallels don't end there. Addiction counselling, in particular, used to work like old-fashioned police interviewing. The 'counsellor' would berate the client for their addiction, with the idea of forcing them to confront it face on. What actually happened, of course, was that the client would push back, storm off or clam up. It was only in the 1980s that two psychologists – Stephen Rollnick and William Miller – came up with what they called 'Motivational Interviewing'. The idea of motivational interviewing is to use trust and empathy – rather than confrontation – to get the client to a point where they not only feel comfortable sharing their story, but actually *want* to.

The Alisons wondered if this approach, which was soon found to be far more effective than traditional counselling, would translate into police interviewing. They developed what they called the 'rapport-based' method, based – as its name implies – around the idea of building rapport with the suspect. This doesn't just mean 'being nice' (as Professor Alison dismissively calls it, 'cappuccino and hugs'), which is likely to come across as false: of course, the suspect knows that you likely don't approve of what they've (allegedly) done. 'Empathy is imagination, not warmth,' Professor Alison told me. It's not about liking the suspect, but imagining what it's like to be them; showing a genuine curious interest in their story, and being open about the fact that they aren't obliged to share it with you. The only fictional interviewer who comes close, Alison told me, is the 1970s TV detective Columbo, with his unassuming manner, battered old Peugeot and trademark curiosity ('Just one more thing . . .'). Look what happens when the interviewer of the terrorism suspect we met above is replaced with an interviewer who uses rapport rather than confrontation:

> Interviewer: 'On the day we arrested you, I believe that you had the intention of killing a British soldier or police officer. I don't know the details of what happened, why you may have felt it needed to happen, or what you wanted to achieve by doing this. Only you know these things [NAME]. If you

are willing, you'll tell me, and if you're not, you won't. I can't force you to tell me – I don't want to force you. I'd like you to help me understand. Would you tell me about what happened? [shows blank notebook] You see? I don't even have a list of questions.'

Suspect: 'That is beautiful. Because you have treated me with consideration and respect, yes I will tell you now. But only to help you understand what is really happening in this country.'

Anecdotes are all very well, but what do the actual data say? In 2013, the Alisons and their collaborators published a paper summarizing years of painstaking work.[9] The team watched 418 video interviews of terrorist suspects who were eventually found guilty, and coded them – sentence by sentence – on the interviewer's technique. Were they confrontational (*judgemental, sarcastic, dogmatic, patronizing, formulaic, distrustful, attacking*) or rapport-building (*non-judgemental, respectful, confident, supportive, humble, trustful, warm*)? They also coded the suspects' responses for 'interview yield': information about capability (does the suspect have the necessary ability to commit the crime? For example, bomb-making skills), motive (a reason to commit the crime) and opportunity (were they in the right place at the right time?); information about who else was involved, where and when. How many pieces of information did the suspect give that are potentially useful evidence?

The results were clear: the more interviewers used rapport-building techniques, the higher the 'interview yield' from the suspect. What is more, the Alisons and their team found that even a small amount of 'maladaptive interrogator behaviour' – that is, losing your rag just once or twice – significantly reduced this yield. The lesson is that setting yourself up as a hero doing battle with an evil monster is setting yourself up for failure. Success comes only when you avoid 'monstering' the suspect (and remember, all were subsequently found guilty of terrible crimes) and building a connection with them as a fellow human.

In the criminal justice system, then, the Monster framing is bad news. And when I say it's bad news, I mean that it literally costs lives. What the rather dry-and-dusty sounding 'interview yield' hides is the fact that if you, as interviewer, 'monster' the suspect, you're decreasing the chances of getting information that could ultimately lead to their being convicted and locked up, and increasing the chances they'll be let out to kill again.

PLOT TWISTED

But what about when we're thinking of a whole different category of evil: Hitler and the Nazis? Just like the fictional monsters we met at the start of this chapter, they killed millions without remorse, were utterly treacherous and deceitful (witness, for example, Hitler's broken agreements with Britain's Neville Chamberlain and the USSR's Joseph Stalin), terrorized whole populations and threatened the way of life of millions worldwide. And all the while, Hitler – like countless other dictators – was using the Monster masterplot recipe to portray himself and his supporters as heroes, and their entirely innocent victims as monsters. Just how did Hitler and the Nazis get away with such a twisted perversion of this masterplot?

A useful starting point when looking for answers to this question is Stanley Milgram's famous 'Behavioural study of obedience', a staple psychology textbook since the 1960s.[10]

> From 1933–45, millions of innocent persons were systematically slaughtered on command. Gas chambers were built, death camps were guarded; daily quotas of corpses were produced with the same efficiency as the manufacture of appliances. These inhumane policies may have originated in the mind of a single person, but they could only be carried out on a massive scale if a very large number of persons obeyed orders.[11]

Milgram was concerned primarily with understanding the motivations not of Hitler and the Nazi high command, but of the thousands of low- and middle-ranking SS officers who carried out their orders. The question was particularly current in 1963, the year Milgram's study was published: in the same year, philosopher Hannah Arendt published her controversial book *Eichmann in Jerusalem: A Report on the Banality of Evil*, an analysis of the trial of Adolf Eichmann – the Nazi officer responsible for organizing the transportation of Jews to death camps – which had taken place in Jerusalem in 1961. Eichmann claimed to be 'just following orders', an attempted defence that – even if taken at face value – was incompatible with Israel's 1957 'black flag' law, which requires soldiers to refuse to carry out a 'manifestly illegal order [that] should fly, like a black flag, a warning saying: "Prohibited! Illegality that pierces the eye and revolts the heart".'[12] Arendt's analysis – which coined the term 'banality of evil' – was controversial for its assertion that Eichmann 'never realised what he was doing' because he was unable 'to think from the standpoint of somebody else'.[13] 'The deeds were monstrous,' wrote Arendt in a 1971 follow-up piece, 'but the doer – at least the very effective one now on trial – was quite ordinary, commonplace and neither demonic nor monstrous'.[14] Many people were understandably outraged; and, as we will see in more detail later, more recent research suggests that Eichmann not only knew what he was doing, but took twisted pride in his work.

The purpose of Milgram's famous study, then, was to investigate experimentally whether this 'just following orders' defence had any possible merit. Would perfectly ordinary people, who had no reason to bear any malice whatsoever towards the would-be victims, cause them harm, in the form of painful – and possibly even fatal – electric shocks?

The answer – as I'm sure I don't need to tell you – is 'yes, they would'. Milgram's experiment is a mainstay not just of psychology textbooks, but of culture at large. When asked to give painful electric shocks to an (imaginary) participant as part of an (imaginary) study of learning and memory, *every single one* complied, up to a level labelled on the (fake) shock machine as '300 Volts: Intense Shock'. A handful dropped out as the voltage was gradually increased past 360 ('Extreme Intensity Shock') and 420 ('Danger: Severe Shock'), but a majority of participants (26 of the 40) continued right up to the maximum voltage of 450 (marked simply as 'XXX').

Now, most psychology textbooks exaggerate a bit when it comes to Milgram's study. While many will tell you that the participants believed the shocks were lethal, they were clearly told beforehand that 'Although the shocks can be extremely painful, they cause no permanent tissue damage', and reminded of this if they asked. Furthermore, two new analyses of questionnaires and interviews conducted at the time – one published in 2017, one in 2019 – found that many of the participants who went up to the highest shock level didn't fully believe the cover story (or, at least, claimed not to afterwards).[15] So it is certainly *not* true – as many textbooks would have you believe – that participants in a psychology experiment were willing to execute each other in cold blood (how plausible is that, really?). What *is* true, however, is that participants were perfectly willing to give each other extremely painful shocks. We know this thanks to a 2016 replication of Milgram's study which used real shocks.[16] And, just to ensure that everyone knew the shocks were real, the participants tried them out for themselves in a 'calibration phase' of the experiment before the main study began.

Where most psychology textbooks, and other popular retellings, *really* get it wrong though is in their *interpretation* of Milgram's findings. The explanation we are usually offered is that the majority of participants were willing to give painful (or possibly even lethal) shocks, simply because someone in authority (or sometimes 'someone in a white coat') told them to. The implication of this interpretation is that the 'just following orders' defence might just hold water: perfectly nice and normal people will do bad things just because they blindly follow the orders of an authority figure.

Importantly, though, this is not the explanation that Milgram himself gives. His interpretation of the findings is much more nuanced:

> The experiment is, on the face of it, designed to attain a worthy purpose – advancement of knowledge about learning and memory. Obedience occurs not as an end in itself, but as an instrumental element in a situation that the subject construes as significant and meaningful. He may not be able to see its full significance, but he may properly assume that the experimenter does.
>
> The subjects are assured that the shocks administered to the subject are 'painful but not dangerous.' Thus they assume that the discomfort caused the victim is momentary, while the scientific gains resulting from the experiment are enduring.
>
> The experiment is sponsored by and takes place on the grounds of an institution of unimpeachable reputation, Yale University.

It may be reasonably presumed that the personnel are competent and reputable. The importance of this background authority is now being studied by conducting a series of experiments outside of New Haven, and without any visible ties to the university.[17]

These later experiments, conducted in downtown Bridgeport, Connecticut, found that – as expected – participants showed much lower rates of compliance. As someone who has visited both Bridgeport and Yale (while working at a Connecticut summer camp), I can confirm that they are the very epitome of chalk and cheese.

The implication is clear: in the original Yale studies, participants showed high levels of compliance because they bought into the cover story and genuinely believed that the 'learning and memory experiment' served an important scientific purpose. OK, so they might have only the dimmest understanding of that purpose themselves, but if a professor at – of all places – Yale ensures them that the study serves an important scientific purpose, then it probably does. In this interpretation, 'compliance' isn't quite the right word for what the participants were doing. Far from merely going along with something against their better judgement 'just because somebody told them to', they were continuing to take part – at great personal discomfort to themselves – in an enterprise whose goals they firmly believed in.

In Milgram's experiment, then, participants displayed monstrous behaviour not because they had become 'evil', but because – in their own minds – they wanted to be 'good' participants; to carry out the experiment for which they had signed up (and for which they were being paid) exactly as the experimenter wanted them to, in order to serve the higher purpose of furthering science.[18]

Returning to Eichmann, the implication is that he and the other middle-ranking Nazis were following orders not merely because they had been told to, but – far more monstrously – because they agreed with and supported racist Nazi ideology. Of course, it is always difficult to generalize from laboratory studies to the real world. But in this case, we don't have to. A detailed study of Eichmann published in 2004 found that beneath his placid and bureaucratic exterior, he was an enthusiastic supporter of Nazism.[19] Before his trial, he expressed regret not for killing millions of Jews, but for the fact that he didn't kill more. And when he felt his superiors, including Heinrich Himmler, were not sufficiently committed to the project, he challenged and even disobeyed them. The sick truth is that Eichmann cast himself in a twisted Monster narrative in which he was the hero, and the Jews, the monster.

How could anyone believe such a thing? The answer lies with the power of masterplots, which, as we have seen throughout this book, can be used either for good or – as in this case – for evil. We are all familiar with the racist propaganda that the Nazis spread long before they started to implement their monstrous 'final solution'. What the case of Eichmann and others like him teaches us is that this racist propaganda was not just a kind of prelude to the horrors that followed or – compared to the death camps – a case of mere insults or name calling. The Nazis' racist propaganda was the foundation of the whole horrific enterprise. The death machine was able to function only because Hitler had sold everyone, including himself, a twisted, upside-down narrative in which the Nazis were the heroes and the Jews, and the other minority groups they targeted with mass murder, were the monsters. As Julia Shaw notes in *Evil*, Nazi prison guards were encouraged to see themselves as 'heroes' for enduring all the horrors they had to witness.[20] On the 'monster' side, the Nazis' anti-Jewish propaganda followed, to the letter, the template for a literary 'monster'. The Jews were portrayed as pure evil, the embodiment of selfishness, an existential threat to the entire nation and its very way of life, as a perversion of the human form – subhuman 'Untermenschen' – or as not humans at all, but vermin or insects. The Jews were portrayed as devious and untrustworthy, as having shadowy, almost mystical powers. A propaganda newspaper, *Der Stürmer*, even claimed that Jews kidnapped and killed Christian children in order to use their blood in religious rituals. A children's book called *Der Giftpilz* portrayed the Jewish race as a poisonous mushroom, as 'The Devil in human form'.

'Controlling the narrative' is a modern – and sometimes mocked – buzz-phrase, but the Nazis understood that controlling the narrative is everything. Rank-and-file officers, and many ordinary Germans, went along with the Nazis, not because they were acting unthinkingly or 'just following orders', but because they had somehow been persuaded by the narrative that the Nazis were selling. The Monster narrative is an extremely powerful one; if we are to prevent atrocities in the modern era, we must be vigilant to any attempts to portray certain groups as 'heroes', and others as 'monsters' for the supposed heroes to overcome.

HAPPY ENDINGS

In a Monster story, an unlikely hero – a stand-in for the reader who represents everything that is noble about humanity – vanquishes a monstrous monster: a villain who is not just devious, cunning, treacherous and slippery, but literally inhuman. This monster is threatening not just our hero, but the whole town, kingdom or species; the hero isn't just trying to save his own skin, but going to war on behalf of all of us. It's just as well, then, that the hero eventually – but inevitably – triumphs over the monster, almost always using some kind of magic weapon, provided by a mysterious armourer. This leaves the hero free to enjoy his reward – usually the hand in marriage of the most eligible bachelorette in town. That's the fictional version, but when applied to real life, the Monster masterplot can serve as a catalyst for human progress in three quite distinct ways.

First, on a positive note, adopting a Monster framing can be useful when doing battle with addiction. As we saw with Claudia Christian's story, this narrative avoids blaming the victim, and instead rightly casts the addiction as the monster, and the victim as the hero who must vanquish it. It also allows us to see that, just as in fictional monster stories, the hero can't triumph by sheer courage or might alone; they need a magical weapon, in Christian's case, the naltrexone injections that really did slay the monster instantly, as if by magic.

In most cases, though, the Monster masterplot serves as a catalyst for human progress by showing us what *not* to do (in this respect, Monster – alongside Icarus – is an exception amongst the masterplot recipes). Whether you're a detective interviewing a terrorist suspect or – more likely for most of us – a parent arguing with a badly behaved teenager, the temptation to apply the Monster masterplot is overwhelming. The lesson we have learned in this chapter, though, is that it is counterproductive. Whether your goal is to extract useful information from a criminal suspect or to improve your teenager's behaviour, setting up a Monster narrative in which you are the hero and your opposite number is a villain, a monstrous monster, is setting yourself up to fail. It is never easy, and for police interviewers it can take months of training, but success will come only if you are able to set aside the Monster masterplot, and see things from the other person's point of view. You don't have to *understand* their motivations, much less *agree with* them; but you have to be curious. Why, in their mind, are *they* the hero and *you* the monster?

MONSTER

The third and final way that the Monster masterplot can act as a catalyst for human progress is by serving as a cautionary tale, a red flag. We must always be on our guard for situations in which bad actors are seeking to frame innocent people as monsters, such as when the Nazis portrayed the Jews as 'Untermenschen', or the devil in human form, killing Christian children to use their blood in monstrous rituals. Alarmingly, this kind of right-wing rhetoric is on the march worldwide. In the decades immediately following the Second World War, it would have been unthinkable for a UK prime minister to refer to immigrants as a 'swarm' (as David Cameron did in 2015)[21] or for a leading US presidential candidate (Donald Trump in 2023) to talk about immigrants 'poisoning the blood of our country'[22] or to refer to his opponents as 'vermin':[23] yet here we are. It's easy to dismiss this kind of thing as just an unfortunate choice of words, but the Monster masterplot recipe shows us why it is so dangerous. It's not just name-calling; it's framing the narrative. Once you have bought into a narrative under which – to put it bluntly – white Americans are the heroes and immigrants the monsters, it's a short and, in their own minds, logical step for these 'heroes' to arm themselves and go to war against the 'monsters'. This isn't fanciful speculation, it's already happening, with armed vigilante groups such as the Proud Boys, Oathkeepers and Arizona Border Recon patrolling the US–Mexican border.[24] Humanitarian groups have seen their water stations sabotaged which, for migrants lost in the border deserts, could be a death sentence; but the perverted logic makes sense under a Monster framing: the hero's goal is to slay the monster, not give him a drink. The lesson of the Nazis, then, shows us why we need to be on our guard against this kind of Monster narrative framing, and challenge it strongly whenever and wherever it raises its ugly head.

6.

FEUD

Brothers Charles and Adam Trask are opposites. Charles is rash and violent, and craves his father's love and respect. Adam is kind, gentle and trusting to a fault, though has little love for his overbearing father. After travelling the country, first with the army then as a hobo, Adam returns to the family farm to learn from his brother that their father – who had been working as a military advisor in Washington DC – has died, leaving the brothers a considerable fortune. Charles, almost literally, ploughs his money into the farm. But Adam sets off for California with his new wife Cathy, a scheming trickster who is always on the lookout for her next mark. Before they leave, Cathy engineers a one-night stand with Charles, despite the fact that he distrusts and dislikes her, recognizing in her his own character flaws.

Flush with money, Adam buys the best farm he can find and sets about building his Eden for his now-pregnant wife, completely oblivious – even though she tells him flat out – that she plans to leave right after the birth. Sure enough, before the babies – twins! – have even been named, Cathy flees, shooting Adam in the shoulder when he tries to stop her. In the big city, Cathy becomes a prostitute, ingratiates herself with the madam to the point that she is named her honorary daughter and

The third and final way that the Monster masterplot can act as a catalyst for human progress is by serving as a cautionary tale, a red flag. We must always be on our guard for situations in which bad actors are seeking to frame innocent people as monsters, such as when the Nazis portrayed the Jews as 'Untermenschen', or the devil in human form, killing Christian children to use their blood in monstrous rituals. Alarmingly, this kind of right-wing rhetoric is on the march worldwide. In the decades immediately following the Second World War, it would have been unthinkable for a UK prime minister to refer to immigrants as a 'swarm' (as David Cameron did in 2015)[21] or for a leading US presidential candidate (Donald Trump in 2023) to talk about immigrants 'poisoning the blood of our country'[22] or to refer to his opponents as 'vermin':[23] yet here we are. It's easy to dismiss this kind of thing as just an unfortunate choice of words, but the Monster masterplot recipe shows us why it is so dangerous. It's not just name-calling; it's framing the narrative. Once you have bought into a narrative under which – to put it bluntly – white Americans are the heroes and immigrants the monsters, it's a short and, in their own minds, logical step for these 'heroes' to arm themselves and go to war against the 'monsters'. This isn't fanciful speculation, it's already happening, with armed vigilante groups such as the Proud Boys, Oathkeepers and Arizona Border Recon patrolling the US–Mexican border.[24] Humanitarian groups have seen their water stations sabotaged which, for migrants lost in the border deserts, could be a death sentence; but the perverted logic makes sense under a Monster framing: the hero's goal is to slay the monster, not give him a drink. The lesson of the Nazis, then, shows us why we need to be on our guard against this kind of Monster narrative framing, and challenge it strongly whenever and wherever it raises its ugly head.

6.

FEUD

Brothers Charles and Adam Trask are opposites. Charles is rash and violent, and craves his father's love and respect. Adam is kind, gentle and trusting to a fault, though has little love for his overbearing father. After travelling the country, first with the army then as a hobo, Adam returns to the family farm to learn from his brother that their father – who had been working as a military advisor in Washington DC – has died, leaving the brothers a considerable fortune. Charles, almost literally, ploughs his money into the farm. But Adam sets off for California with his new wife Cathy, a scheming trickster who is always on the lookout for her next mark. Before they leave, Cathy engineers a one-night stand with Charles, despite the fact that he distrusts and dislikes her, recognizing in her his own character flaws.

Flush with money, Adam buys the best farm he can find and sets about building his Eden for his now-pregnant wife, completely oblivious – even though she tells him flat out – that she plans to leave right after the birth. Sure enough, before the babies – twins! – have even been named, Cathy flees, shooting Adam in the shoulder when he tries to stop her. In the big city, Cathy becomes a prostitute, ingratiates herself with the madam to the point that she is named her honorary daughter and real-life heiress, then poisons her. Adam falls into a deep depression and doesn't even name the boys, until he is eventually forced to by Samuel Hamilton – a friend from a neighbouring, but much poorer, farm – with the aid of Adam's Chinese servant Lee. Having decided against calling them Cain and Abel, which they think would be tempting fate, they call them – um – Cal and Aron. We never get to find out whether the boys' biological father is Adam or Charles (with whom Cathy had a one-night stand, remember).* Are you keeping up at the back? I hope so, because our story hasn't even begun yet – that was all back story, told in flashback.

The real story is that of Aron and Cal. Aron is like Adam (and the biblical Abel) – kind, gentle, trusting and, like his mother, blonde and beautiful. Cal is like Charles (and the biblical Cain) – rash, violent, dark and, like his mother, scheming and cruel. Aron falls in love with a girl called Abra, but the relationship is doomed by the fact that he sees her as something of a replacement mother. On his deathbed, Samuel Hamilton tells Adam of Cathy's true whereabouts. He visits the brothel and finds Cathy (now rebranded Kate) but, when she rails at him, feels pity rather than hate. He doesn't tell the boys, allowing them to continue to believe that their mother is dead. Eventually, Cal – stumbling around town at night – finds out the truth for himself, but – knowing it would crush his gentle soul – keeps it from Aron.

After Adam blows almost his entire fortune on a doomed project to ship refrigerated lettuce across the country, Cal vows to earn the money back to win his father's respect (which Aron

---

* It's *just about* possible – as some commentators have suggested – that Charles and Adam fathered one twin each, although with only around twenty known cases of 'heteropaternal superfecundation' to date worldwide, it seems unlikely that this is what the author had in mind.

already has, having aced high school and been accepted to Stanford). Partnering up with Samuel Hamilton's son Will, scheming Cal sets up a business that involves buying cheap beans from desperate local farmers and selling them, at an inflated price, to countries in war-torn Europe. The scheme is a success and Cal proudly presents his father with $15,000 in cash when the family are together for Thanksgiving. Adam refuses to accept what he considers to be dirty money and lets it be known that he's proud of Aron, but not Cal. Predictably, Cal takes this badly, and gets his revenge by introducing Aron to their mother at the brothel, which he knows will kill him, almost literally; a despondent Aron signs up for the army, where he will soon die in battle.

It's too late for Aron to forgive Cal, but will their father, Adam? While all of this has been going on, Adam and his good-natured neighbour Samuel have been engaging in long theological discussions with Adam's loyal Chinese servant Lee, who has virtually become part of the family. In particular, they obsess over the true moral of the biblical story of Cain and Abel, sons of Adam and Eve (yes, that Adam and Eve!). Cain, a farmer, and Abel, a shepherd, offer sacrifices to God of vegetables and a lamb, respectively. God accepts Abel's sacrifice but rejects Cain's as poor quality. God can see Cain's temperature rising, and warns him, rather poetically, to keep a cool head ('If you do not do well, sin is lurking at the door; and its desire is for you, but you must master it'). It's no good – in a fit of jealous rage, Cain kills Abel, and God curses Cain to wander the Earth, homeless.

Lee explains that he and his elders have been learning Hebrew, and have come to the conclusion that the usual English translations of the Cain and Abel story are mistranslations: God doesn't say you 'must' master sin (as in the American Standard Bible) or 'shalt' (i.e. will) master sin (as in the King James Bible), but that you *may* master sin. The crucial Hebrew word, according to Lee, is *timshel*.[1] And it is this word that Adam, on his deathbed, whispers to Cal. You're not doomed to follow in the footsteps of your mother, or to be crushed by your guilt. You have a choice, you may reject sin: '"*Timshel!*". His eyes closed and he slept. THE END'.

THE MASTERPLOT RECIPE

John Steinbeck's *East of Eden* (named after the area in which Cain lived after being exiled by God) is the quintessential revenge-and-rivalry Feud story. In fact, according to Steinbeck himself, it's 'perhaps the greatest story of all', 'the first book'.[2] He wasn't being boastful – what he meant was that the Feud masterplot, characterized by revenge and rivalry, sums up the human condition; is 'the basis of all human neurosis'. *East of Eden* epitomizes the Feud masterplot that underlies stories as diverse as *Top Gun* (both the original and the 2022 reboot), *Mean Girls* (both the original and the 2024 reboot), *The Godfather* (especially *Part II*), *Crime and Punishment*, *The End of the Affair*, *Frozen*, *The Count of Monte Cristo*, *Rocky*, *Better Call Saul* and *Beef*.

The first key ingredient of the revenge and rivalry masterplot is that the two sides have to be more-or-less **evenly matched**. If not, we have a very different masterplot on our hands: David versus Goliath isn't Feud, it's Monster. Cinderella versus her sisters isn't Feud, it's Underdog. But Aron versus Cal, Michael versus Fredo Corleone, Anna versus Elsa or – in the real world – Coke versus Pepsi, are broadly even match-ups.

But this doesn't mean the rivals are *the same*; the feuding pair are never clones or identical twins. The second key ingredient is that the rivals must be **mirror images** – yin and yang – the strengths of the first are the weaknesses of the second, and vice versa. In *East of Eden*, Aron is kind, but fragile and naive. Cal is mean, but strong and cynically clever. In *Top Gun*, Maverick (the Tom Cruise character) is hot-headed, selfish, untrustworthy, a show-off, not a team player. Iceman (Val Kilmer) is cool, generous, reliable, responsible, one of the team. In *Frozen*, Anna is bubbly, happy-go-lucky, scatty and perhaps a bit of a tomboy. Elsa is colder, serious, organized and stereotypically ladylike.

The third key ingredient of the Feud masterplot is that after the rivalry has been established (typically, we meet one or more of the rivals in the first act, before they encounter each other), the feud takes on **a life of its own**, above and beyond what the original quarrel was about. At the heart of Netflix's *Better Call Saul* is a feud between two siblings who – like Cain and Abel, Charles and Adam, Cal and Aron or Anna and Elsa – are chalk and cheese. Chalk – sorry, Chuck – is a well-dressed, highly respected partner at a leading corporate law firm. His brother Jimmy wears cheesy ties, has a mail-order law degree, and – no stranger to petty crime himself – defends low-level offenders. The two start out as friends – Jimmy caring for Chuck when he becomes housebound due a psychosomatic medical condition – and their initial skirmishes are minor but, by the end of Season 3, the two are doing battle in the courtroom, with Jimmy tricking Chuck into launching a tirade that sees him ultimately being forced out of his own law firm. But at no point is there any motivation for either brother's attack beyond retaliation for the other's, with the original quarrel long forgotten. This trope of the feud taking on a life of its own is taken to its logical conclusion in Netflix's *Beef* between down-to-earth handyman Danny Cho and uptight businesswoman Amy Lau. The feud is triggered when Danny and Amy almost – but don't – collide in a car park. That's it. Yet, nine episodes later, and Danny and Amy are trying to frame each other for arson and armed robbery, respectively. None of the minor escalations along the way felt wildly disproportionate but, as in the story of the Canadian blogger who traded his way up from a paperclip to a house, they all add up.[3]

On a more positive note, the fourth and final key ingredient of the Feud masterplot is **reconciliation and redemption**. If this doesn't happen, we are into the realms of the Icarus masterplot. *East of Eden* is somewhat unusual in that Aron is already dead, leaving Adam to forgive Cal on Aron's behalf. But in most cases, this redemption takes the form of the rivals joining forces to take on a bigger and deadlier foe, finding a new – if often grudging – respect for one another in the process. In the climax of *Top Gun*, Maverick shoots down three MiG fighters but is ultimately rescued by Iceman, who shoots down one more, causing the remaining two to retreat. In Disney's *Frozen*, the feuding sisters Elsa (cold, scheming, cunning) and Anna (warm, innocent, naive), reunite to foil Prince Hans's plot to seize the throne of Arendelle. In the final two episodes of *Beef*, Danny and Amy run each other off a cliff and, stranded with no food, water or phone reception, must work together to survive. In the closing scene, Amy visits an unconscious Danny in hospital. She hugs him and, lapsing into consciousness, his arm seems to hug her back.

STRANGER THAN FICTION

The 2019 Super Bowl was the lowest scoring ever, with the New England Patriots becoming only the second team in history to win the title while scoring just a single touchdown. The LA Rams, for their part, were only the second ever Super Bowl team to fail to score a touchdown entirely. In (proper) football terms, it was about as exciting as the 1994 World Cup Final in California, where Brazil beat Italy on penalties after a full two hours in which neither team scored.

The half-time show was even worse. In the wake of the controversy surrounding the NFL's decision to ban players from taking the knee to protest racism, anyone who's anyone had ruled themselves out, leaving Maroon 5 to carry the can (they did manage to rope in Outkast's Big Boi, André 3000 having turned them down).[4]

No, the real entertainment was to be found on the walls, billboards and recycling bins of downtown Atlanta, and on TV screens around the country, where a hundred-year-old battle was playing out. Atlanta, you see, is the home of Coca-Cola, whose rival Pepsi had long been one of the main Super Bowl sponsors. Pepsi couldn't resist the opportunity to taunt its big brother, plastering most of Atlanta with adverts reading 'LOOK WHO'S IN TOWN FOR SUPERBOWL LIII', 'PEPSI IN ATLANTA. HOW REFRESHING' and, perhaps most snarkily of all, 'HEY ATLANTA, THANKS FOR HOSTING. WE'LL BRING THE DRINKS'.[*]

Coke versus Pepsi, of course, is the nonpareil of the Feud masterplot, with all of the key ingredients present in spades. **Evenly matched?** OK, so Coke has always had the edge in terms of market share, but in terms of the product itself, nobody can tell the damn difference. That's right – even you. In the course

---

[*] In terms of TV ads, Coca-Cola went for an aspartame-sweet animation, 'A Coke is a Coke', which referenced its 1970s one-big-happy-family 'I'd Like to Teach the World to Sing' campaign, the thrust of which was 'different is beautiful', but we can all enjoy a Coke. Pepsi ran an oddly defensive ad in which a customer orders a Coke, is asked 'Is Pepsi OK?', and thereby triggers a minute-long Steve Carell rant ('Are Puppies OK? Is the laughter of a small child OK?') while Lil Jon and Cardi B strut their stuff.

of promoting my first book, *Psy-Q*, I ran a blind taste test challenge with volunteers at venues ranging from the Royal Institution in London, to Google's UK HQ, to a sixth-form college in Hull. The volunteers put on a blindfold, I put a sanitized metal straw into a can of each beverage, served them to the victim in a randomized order and asked them to guess Coke or Pepsi on each sip. I did this ten times per person. To beat chance to the satisfaction of a statistician (this was all done to prove some arcane point about statistics), you have to get 9/10 correct. *Nobody has ever done it.* In fact, more often than not, I ended up stopping the test halfway through, because the cocky volunteer – I've never failed to find a volunteer who isn't *sure* they can do it – had already got two wrong. And it's not just me; a 2004 study in the prestigious journal *Neuron* found not only that people couldn't tell the difference, but that there was *no relationship at all* between which of the two drinks people *said* they preferred and which they *actually* preferred in a blind taste test.[5]

Mirror images? Absolutely. Coca-Cola, Diet Coke and Coke Zero are mirrored by Pepsi, Diet Pepsi and Pepsi Max; Coca-Cola Cherry by Pepsi *Wild* Cherry; Fanta, Sprite, Powerade and Monster by Crush/Tango, Starry/7-Up, Gatorade and Rockstar. But it's not just mere equivalence. Again, like Iceman and Maverick, the rivals have mirroring brand identities. Coke is traditional, family friendly, authentic ('The real thing'), even conservative. Pepsi is brash, irreverent, young ('The choice of a new generation'), even anti-establishment. The Michael Jackson advert that launched the new-generation slogan was quite radical for its time (1984) in featuring an almost all-black cast.

Feud taking on **a life of its own**? To the (Pepsi) max! This is particularly true for consumers who, remember, can't tell the difference between the two companies' flagship beverages in a blind test. In fact, the 2004 *Neuron* study went even further in demonstrating that what it rather snootily calls our 'Behavioral Preference for Culturally Familiar Drinks' (translation: 'which cola you like') is really all in the mind. As well as just asking participants which drink they preferred and conducting the standard blind taste test, the experimenters hooked participants up to an fMRI brain scanner, before conducting a taste test with a twist. The participants – unbeknown to them – actually got Coke every time, but sometimes served in a Coke-branded cup, sometimes a plain, unbranded cup. In terms of both what participants *said* they preferred, and what their brain scans said they *actually* preferred, the Coke tasted better when it was given in the branded rather than unbranded cup. Fascinatingly, the same was *not* true for Pepsi, which tasted just as good in an unbranded as branded cup. These findings show that the Coke-vs-Pepsi rivalry has taken on a life of its own which has absolutely nothing to do with the relative merits of the stuff in the can, and absolutely everything to do with the branding around it. But it's also true for the companies themselves. Pepsi's 2019 Super Bowl billboard campaign said nothing about the merits of its product; it trolled Coke all over Atlanta for sheer banter. The same is true for Pepsi's 2020 Halloween ad showing a can of Pepsi in a red 'Cola Coca' cape (gotta watch out for those trademark lawyers!) with the line 'We wish you a scary Halloween!'

What about the final ingredient of **reconciliation and redemption**: will Coke and Pepsi ever put aside their differences to fight a common enemy? We're halfway there. Forty years ago, the adverts were brutal. 'Coca Cola *says* it's the real thing' sneers a Pepsi advert from 1982, showing a close-up of a Coke can, before introducing the Pepsi challenge, the conclusion of which is flashed onscreen: 'Nationwide more people prefer the taste of Pepsi over Coca-Cola'. These days, Pepsi's ribbing of Coca-Cola is just good-natured banter, with no attempt to convince drinkers that its offering is actually superior. And how did Coke respond to the Super Bowl trolling? With nothing but magnanimity: 'We're thrilled to help our city welcome everyone to town for the Big Game, including our friends from Pepsi'.

Partly, this is just good marketing. Counter to what most people have long believed, a 1999 review and meta-analysis – albeit one conducted in the domain of political adverts – found that so-called negative advertising – attacking a rival – is not particularly effective.[6] Neither would it seem wise to feature a rival's product, and even its slogan, as prominently as in Pepsi's 'Coca-Cola *says* it's the real thing' ad.

Partly, somewhat more sinister motives seem to be at play. Around the world, more than thirty different jurisdictions, including the UK and some parts of California, already levy additional taxes on drinks with high sugar content.[7] *Yes! To Affordable Groceries*, which describes itself as 'a group of citizens, businesses and community organizations', campaigns against these kinds of taxes. But, surprise, surprise, historically its biggest funders have been Coca Cola ($3.8 million in 2018) and Pepsi ($2.8 million), with Dr Pepper (around $1 million) and Red Bull ($100,000) also on board.[8] That's right, in classic Feud style, just like Anna and Elsa, the long-time rivals have teamed up to fight an even greater foe: legislation that threatens their profit margins.

THE SCIENCE BEHIND THE STORY

Perhaps more so than for any other masterplot, the seeds of Feud have already taken deep root within us by childhood. In fact, they may even have been planted before birth; hardwired into our DNA.

THE STORIES OF YOUR LIFE

In 1970, the British psychologist Henry Tajfel published a landmark study designed to investigate the roots of conflict between groups.[9] Tajfel starts his paper by discussing the hostility shown towards Bosnians in parts of what was then Yugoslavia (which boiled over into war in the 1990s), and towards immigrants to the UK from the West Indies, Pakistan and India (Tajfel himself was an immigrant, originally from Poland). He discusses racial conflict in the US, religious conflict in Northern Ireland and – somewhat bizarrely to modern ears – linguistic national conflict in Belgium. Tajfel's theory was that while these conflicts might on the surface *seem* to be about race, or religion, or language, in fact they are not really *about* anything at all. Rather, whenever two or more groups emerge – however they happen to be defined – conflict between them is almost inevitable. Tajfel didn't frame his theory in terms of narrative per se, but it is clearly built around what we have identified above as a key ingredient of the Feud masterplot: the fact that the feud seems to take **on a life of its own,** above and beyond whatever the original disagreement was about.

To put his theory to the test, Tajfel recruited a group of Bristol schoolchildren and split them into two groups on the basis of the most trivial thing he could think of. He showed them slides of abstract paintings by Paul Klee and Wassily Kandinsky, and assigned them to groups on the basis of which they preferred. That's all. He didn't – as in the now largely debunked 'Robber's Cave' experiment from the 1950s – send the boys off to camp in the woods and engineer conflicts between them.[10] He didn't need to. The mere fact of creating two arbitrary groups was enough.

In the main part of the experiment, the boys were given the task of allocating points to one another; points that could be translated into money afterwards.* Ostensibly, the points were a reward for some trivial dot-counting task, but this was just a cover story. What Tajfel was really interested in was how they allocated the points. Each boy was given the task of awarding points to two other boys: one from his own group; one from the opposite group. That is, a boy on Team Klee would be asked to award points to another boy from Team Klee, and to a boy from Team Kandinsky. The awardees were anonymous; the boy giving out the points didn't know anything about them, other than their preferences in abstract art. Crucially, the boys couldn't just give out the points willy-nilly – on each run, they had to choose from a set of options: (a) 8 points for your teammate, 7 points for the boy from the other team; (b) 9 points for your teammate, 14 points for the boy from the other team, and so on. So how did the boys dish out the points? What would you do?

One possible strategy is simply to maximize the total number of points that you give out, regardless of how they're shared between your teammate and the boy from the other team. Needless to say, the boys didn't choose that strategy; neither did they choose the strategy of sharing the points out as equally as possible. A third strategy is just to give as many points as possible to your teammate, and forget about the other guy. But the boys didn't do that either. Instead, they reliably chose the spiteful strategy of maximizing the *difference* between the points given to the teammate and those given to the 'opposing' team member. That is, the boys were prepared to reduce the pay-out to their own teammates if, by doing so, they could reduce the pay-out

---

* This method is a distant cousin of the ultimatum game we met earlier, in that there is no opportunity for the receiver of the points to reject the proposed allocation.

to the other team even more. Like petty feuding neighbours, the boys would rather trample on next door's roses than grow their own, all because the other guy happened to put his hand up for a different artist.

At least on the surface. Tajfel's whole point, of course, was that this financial feuding was in fact nothing to do with modern art, and everything to do with spontaneously occurring rivalry between groups. Depressingly, a modern-day twist on Tajfel's study suggests that this tendency is present even amongst fourteen-month-old babies.[11] First, the babies in the study were invited to select a snack from a bowl: either a green bean or a Graham cracker (I don't know either – ask an American). The babies were then introduced to two rabbit puppets. The first identified themselves as 'similar' to the child. For example, if the child chose a Graham cracker, the puppet said, 'Mmm, yum! I like Graham crackers', and 'Ew, yuck! I don't like green beans' (or vice-versa if the child chose a green bean). The second rabbit puppet identified themselves as 'different' to the child, claiming to dislike the food that the baby had picked and to like the other.

In the test phase, these two rabbits each played with a ball. But then – oh no, disaster – they lost control of the ball, and it rolled away to the other side of the screen, where one of two dog puppets was waiting. On some test runs, the waiting dog was a 'helper' dog, who took the ball back to the rabbit who had lost it. On other test runs, the waiting dog was a 'harmer' dog, who looked different to the helper dog, and who pinched the ball and ran off with it. Finally, the two dog puppets were set out on the table, and the infants were invited to choose one to play with. This was taken as a measure of which dog they liked best.

Did these fourteenth-month-olds choose the helper dog or the harmer dog? It depends. Remember how one rabbit was 'similar' to the child (in that they liked the same food) while the other one was 'different'? Not that surprisingly, children preferred the dog who helped the rabbit that was 'similar' to them. But – in a rather sinister twist – these infants also preferred the dog who harmed the rabbit that was 'different' to them.

The unfortunate conclusion of the science, then, is that we don't start out as egalitarian innocents, who are then corrupted by a cruel world. We are born with (or, at least, acquire in the first years and months of life) a preference for those who we see as similar to ourselves. We look favourably on those who help people 'like us' and who harm people 'not like us'. In other words, Steinbeck was right: the Feud masterplot is indeed 'the first book', etched into our DNA.

UNDER THE INFLUENCE

One particularly interesting aspect of the Feud masterplot is that it's one that we've *all* experienced, perhaps more so than for any of the other plots in this book. In *East of Eden*, when a group of characters are debating the meaning of the Cain and Abel story, the servant Lee notes that 'people are interested only in themselves. If a story is not about the hearer he will not listen. And I here make a rule – a great and lasting story is about everyone or it will not last. The strange and foreign is not interesting – only the deeply personal and familiar'.

What Lee is getting at here is that we have *all* had that feeling of being somehow forced to compete with someone. If you have a sibling – and assuming that, unlike Cain, you haven't murdered them – you'll immediately recognize that feeling of being in competition for affection and attention. Indeed, this competition seems to be the reason for the longstanding finding that, on average, children's IQ-test scores decrease, albeit only by a couple of points, for each position in the family. For example, on average the first-born child in a family has an IQ of around 103, the second-born 101 and the third-born 99.[12] One of the reasons for this pattern is competition between siblings for their parents' attention.[13] A first-born child has a period – typically a year or two – in which they can have all of their parents' attention to themselves. The second-born child always has to share their parents' attention with a sibling; the third-born with two siblings, and so on. Sibling rivalry is not only a very real phenomenon, but one with measurable consequences.

As *East of Eden*'s Lee recognizes, though, the Cain and Abel story resonates with all of us; not just those of us who have siblings. At school, you no doubt competed with other children for the attention and praise of your teacher (and, even more viciously, for the friendship of other children). Perhaps you and a rival at work are competing right now for the attention and praise of your boss? Or perhaps, more seriously, you're formally competing for promotion? And if, like most siblings, you're at least sort-of friends, that probably makes it even worse? Or perhaps you had to compete with a love rival for the affections of your current partner? Or worse, perhaps your would-be partner chose someone else, or your former partner left you for someone else (which has a much more bitter sting than a partner leaving and becoming single)?

It's not all bad news, though. Although our internalization of the Feud narrative can lead us into pointless and unnecessary conflicts (as any parent of siblings can observe on a daily basis), it is also responsible for a whole field of human endeavour that, for millions of people, is one of the major sources of meaning in life.

FEUD

Sport is so pervasive that we rarely stop to notice just how bizarre it is. For literally millions of fully-grown adults all around the world, at regular intervals, the thing they care about most in the world – to the point that it moves them to screams of rage and tears of joy – is whether some people wearing red T-shirts kick the ball between some posts (or throw it in a hoop, or touch it on the ground) more often than some people wearing blue T-shirts. It makes sense only when you remember that all sports – but especially sports based on geographically identified teams – are nothing more than a safe way to gorge daily on real-world Feud narratives (a role formerly played by the much-less-safe outlet of war).

Don't try and tell me it's about appreciating exquisite skill or elite performance. That might be the icing on the cake, but it's not what sport is about. If you were to offer a hundred football fans the choice between attending a charity exhibition match featuring Lionel Messi and Cristiano Ronaldo or a cup final between their team and their local rivals, not one would choose the former, no matter how dire the football served up by their own team. In fact, the more dire the football, the more intense the rivalry. These days, Manchester City probably care more about beating Real Madrid than Manchester United. But nothing could be more important to a Burnley fan than beating Blackburn; to a Wolves fan than beating West Brom; or to us Ipswich fans than beating Norwich.

Sporting feuds contain all of the key ingredients. The rivals must be **evenly matched mirror images**. Ipswich's rivals are Norwich, not the geographically-much-closer Colchester, who we tend to have a soft spot for, albeit in a patronizing kind of way. Manchester United's rivals are Manchester City, not Altrincham or Stockport County (and, indeed, in City's wilderness years, United fans tended to focus more on their rivalry with Liverpool). Feud taking on **a life of its own**? Absolutely. Burnley versus Blackburn – probably the longest-running rivalry in professional sport – isn't about cotton-weaving contracts in the 1800s, who won the FA Cup in the early 1900s, or even – with fans of both teams scattered throughout the north-west – about whether you live in Burnley or Blackburn. It's about the rivalry itself, as they say in that part of the world, 'end of'. **Reconciliation and redemption** opportunities are few and far between in professional sport, but rival fans are certainly capable of coming together to fight a bigger enemy when required, such as when fans of Manchester United, Manchester City, Chelsea, Tottenham, Arsenal and Liverpool united to successfully oppose a breakaway European Super League in 2021.

So perfect, in fact, is sports rivalry as a microcosm of this universal human condition that scientists have long used it as a testing ground for understanding the revenge and rivalry that characterizes the Feud.

The 'ultimatum game' isn't much of a game, but it certainly features an ultimatum: 'Take it or leave it'. The game, such as it is, is played between two opponents. The two are given a sum of money to split between them, but there is no negotiation. One of the players – chosen at random – is given the role of 'Proposer', and comes up with a suggested split. The other player, the 'Responder', has a simple choice: accept the offer, in which case the players receive the money as per the agreed-upon split, or reject it, in which case both players leave with nothing. The original point of the ultimatum game was to demonstrate that – contrary to what economists had long assumed – people are not purely 'rational' when it comes to financial matters. That is, they don't set aside all other considerations and simply take whichever course of action leaves them better off in purely monetary terms; those other considerations matter too, and quite a bit.

Suppose that the amount to be split is $10, and the proposer offers a 90/10 split – $9 to herself, $1 to her opponent. The 'rational' course of action is to accept the offer, since a dollar is better than nothing. But I'm sure you don't need me to tell you that responders almost never accept this type of offer. Would you? Of course not. Almost everyone agrees that it's worth 'paying' a dollar for the satisfaction of telling that greedy so-and-so where to get off. That said, everyone has a price. One group of researchers exploited the poverty of poor villagers in India to run a high-stakes ultimatum game in which the amount to be divided was the local equivalent of around $16,000 per pair (of course, the cost to the research project was far lower, since it was deliberately conducted in very low-wage communities).[14] The typical proposal offered was around 90/10, presumably because the proposers guessed that few responders would turn down the equivalent of $1,600 just to spite them; and this is exactly how it turned out, with 23/24 proposals accepted (props to the single participant who did reject the offer!).

That's the extreme version, though. In the 'normal' version, where participants are playing for $10 or so, proposers typically offer a split of around 75/25, an offer that around 75 per cent of responders accept. This makes the ultimatum game a great laboratory for studying Feud in the context of sports. The most notable study of this type, looking at US college football rivalries, was conducted by researchers at the University of Florida (Go Gators!), Wayne State (Go Warriors!), the University of Georgia (Go Bulldogs!) and the University of Illinois (Go, um, Fighting Illini!). Yes, yes, I know that to us Brits, the idea of college football rivalry sounds laughable, with the average university football match in the UK attended by the proverbial two men and a dog. But the US college system boasts *ten* stadiums with a greater capacity than Wembley's 90,000.[15] The record attendance for a college game – in which the Tennessee Volunteers beat the Virginia Tech Hokies in 2016 – was 157,000. US college football is a BIG deal.

The US research team ran a version of the ultimatum game in which the proposer and responder were either both fans of the same college football team, or of deadly rivals. Perhaps unsurprisingly, proposers offered almost 10 per cent less to fans of the rival team than to fans of their own team, even though they knew nothing else about the other person whatsoever. More surprisingly, though, responders showed a very similar pattern. Even when holding constant the amount they were offered, responders were around 7 per cent more likely to reject *exactly the same offer* when it came from a supporter of the rival team. Throwing traditional economic dogma out of the window, people were prepared to make themselves worse off financially purely to spite a rival fan. It almost goes without saying that the rivalry is misplaced here, in the sense that the ultimatum game is purely about money, and has nothing to do with (American) football.

In a deeper sense, though, the rivalry is not misplaced at all: sports-based rivalries ultimately have nothing to do with the sport itself, and everything to do with the group-based rivalry. This point is neatly illustrated by a 2023 study in which politicians were invited to play the ultimatum game against either a member of their own political party or a rival party.[16] The results were almost identical to the college football study, with both offers and acceptance rates around 8 per cent lower for rivals than (here) party-mates, regardless of whether the politicians were from Belgium, Canada, Germany, Switzerland or the USA. Again, the conclusion is the same: whether the groups are defined in terms of football fandom or political affiliation is immaterial – the Feud takes on a life of its own.

There is hope, though. If sporting and political rivalries are just narratives, then maybe if we can change the narrative, we can change the behaviour. In a classic study from 2005,[17] Manchester United fans were asked to visit the psychology department at Lancaster University and fill in various questionnaires about their fanhood. On their way to a different building, ostensibly to take part in the next phase of the study, each Manchester United fan witnessed a staged accident. A jogger slipped, fell, shouted and lay on the ground clutching his ankle in apparent agony. Would the Man Utd fan play Good Samaritan and go to his aid? The clever part, as I'm sure you've guessed by now, is that in some cases the stricken jogger was wearing a Man Utd shirt; in some cases the shirt of their biggest rivals, Liverpool. And, yes, as you'd expect, these Man Utd-supporting participants almost always helped the Man Utd fan (on twelve out of thirteen occasions, in fact), but helped the Liverpool fan only 30 per cent of the time.

So far, so depressing. But then the experimenters ran a second study. This time, instead of being asked to fill in questionnaires about being a Man Utd fan, the participants were asked to fill in questionnaires about being a football fan. Everything else about the experiment was the same. All the participants were again Man Utd fans, and the jogger again wore either a Man Utd shirt or a Liverpool shirt. All that changed was the narrative framing: 'football fan' rather than 'Man Utd fan'. And I know this sounds too good to be true, but – yes – in this version of the experiment, these 'football fans' were almost exactly as likely to go to the aid of the (apparently) Liverpool-supporting as the Man Utd-supporting jogger. Changing the narrative really does change human behaviour.

PLOT TWISTED

As one of the most powerful and most universal masterplots, the Feud is ripe for abuse and misuse. Given that even something as trivial as liking a different artist, snack or football team is enough to trigger negative behaviour towards others, it's not surprising that insulting or killing a group or family member can lead to feuds that are not just life-long, but can even span multiple generations.

To try to understand more about these real-world feuds, I spoke to Professor Simon Harding, Director of the National Centre for Gang Research at the University of West London. With his neatly trimmed beard, narrow spectacles and gentle Scottish Highlands accent, Professor Harding comes across as the opposite of a typical gang member; something that often works to his advantage. Gang members are happy to talk to him precisely because, as he put it, 'I'm not part of their world'; there's no way that whatever they tell him could get back to other gang members (or, worse, to members of rival gangs). Neither is he averse to playing up his middle-classness in order to get gang members to spill the beans. 'If I want them to go into more detail,' he told me, 'I'll polish my glasses and say [affects hoity-toity professor accent] "I'm *sorry*, I don't know *quite* what you *mean*."'

But Harding is no ingenu; a veteran of the illegal party scene in late 1980s Manchester, he hung out in places like the Kitchen – a nightclub-cum-squat in the notorious Hulme Crescents – and of course the Haçienda; 'Membership number HAC36,' he told me proudly. 'Tony Wilson gave it to me himself.' Harding's experience – he estimates that he's conducted over 5,000 interviews – is valuable when it comes to one of the hardest parts:

getting the interviews in the first place. Obviously, you just can't go and hang out on street corners and expect gang members to chat to you. He has to pay intermediaries, and sometimes the gang members themselves, almost always out of his own pocket (as you might expect, government research funding agencies are queasy about indirectly funding gangs). Harding's most expensive interview cost him £660, as his subject – a notorious acid-thrower – agreed to talk only when offered a free bar in a nearby hotel (a canny move by Harding to ensure that the entry and exit points were covered by CCTV, 'just in case'). It was worth it, though, when the subject offered up the immortal line 'torture in a bottle', which got Harding's research into the headlines worldwide. These interviews were the source material for his magnum opus, *The Street Casino*, a surprisingly accessible and entertaining read for a work of academic criminology.

What really struck me, when talking to Harding and reading his book, was the extent to which – though money plays a role too – 'the game' is fought over respect, reputation and status (grouped together under the term 'street-capital'). When you think about it, middle-class people are really no different; it's just that they derive their respect, reputation and status from their job, their house, the craft beer they drink and – for the truly middle-aged, middle-class male – the bike they ride. Aghast middle-class commentators who describe street violence as 'random' and 'senseless' are missing the point. Just as with middle-class status games, the 'street casino' is run according to strict rules. As one of Harding's interviewees put it:

> You learn it just by hanging around. You know what honour is and what isn't. You know what you are supposed to do and not supposed to do, really. Just like what you are allowed to do at school. They don't even need to explain it.[18]

When an honour code has existed for generations, it is often written down and formalized. One particularly notorious and well-studied example is the Albanian Kanun: a set of ancient tribal laws that – while officially abolished under communism – still hold sway amongst some to this day. Central to the Kanun is the concept of *koka për kokë*, 'a head for a head'. If someone from a different family kills a member of yours, you're not just *allowed* to avenge them, but *obliged* to. On pain of becoming a pariah, you must seek *gjakmarrja*, 'blood vengeance', by killing a male member of the other family (the Kanun is deeply sexist – women are not 'spared' as such, it's more that they are not considered 'worthy' targets). Once you are a target of *gjakmarrja*, you will remain one for life. Your only options are to flee the country altogether or to never leave your house (the code forbids killing a man in his own home). It's the same with the unwritten code of Harding's street gangs. If someone 'tests' you, even in the most minor way, you have to respond. Harding puts it like this:

> An example is the 'visual bump', whereby someone stares at another person for too long, as if sizing them up for a fight (known as 'screwing' in SW9). This is . . . a challenge to both authority and street capital, demanding an immediate response. A failure to respond is tantamount to 'falling', and is seen as an open invitation to victimisation.[19]

A more serious test of course demands a more serious response. As under the Kanun, the code of London street gangs allows targeting family members but – going further than the Kanun – does not exclude women: 'I can't get you, so I got her. But I'll get you as well. I'll get you but I can't see you. But I saw your sister so I mashed her up.'[20]

These honour-based codes might seem barbaric, but in times and places where there is no effective government, they can play an important role in deterring violence. Suppose you are thinking about murdering a rival in ancient rural Albania, which has no police officers to arrest you, no detectives to investigate and no court in which to try you. Your intended victim has no siblings, and his father died years ago, leaving just his eighty-five-year-old mother. What's *she* gonna do? You might as well just go ahead and kill him. But suppose now that you live under the Kanun, meaning that your victim's distant cousins would be obliged to take revenge on you or your family. This knowledge would likely give you pause. Notice that the *obligatoriness* of the revenge is the part doing the heavy lifting here. If your would-be victim's distant cousins simply had the *option* to take revenge on you, they may well not bother. Maybe they didn't like the guy either; maybe they've never even met him. The fact that they *have* to – and almost certainly will – is what holds you back.

It's the same for London street gangs where, technically, the police and the courts have jurisdiction, but – in practice – the prohibition on 'grassing' makes them ineffective. A member of a rival gang won't hold back from initiating a violent attack because he fears arrest and conviction – the chances of which are minimal – but he might well hold back for fear of retribution ('Y'know, don't mess about with him, he'll throw a petrol bomb through your window').[21] In this way, the Kanun and similar honour codes follow the same underlying logic as the Cold War theory of Mutually Assured Destruction (MAD): the enemy won't destroy you if he knows for sure that you will destroy him in return. MAD was taken to its logical conclusion in Stanley Kubrick's *Dr. Strangelove or: How I Learned to Stop Worrying and Love the Bomb*. The US President learns that the Soviets have built a Doomsday Device which, upon detecting a nuclear attack, detonates enough bombs to make the entire planet uninhabitable for ninety-three years.*

Dr. Strangelove – and the entire world – ends when the Doomsday Device is triggered by an American attack, ordered by the insane General Ripper. This illustrates the drawback of honour-based systems like the Kanun. While they can be an effective deterrent to violence most of the time, it takes only one rogue actor to trigger a spiral of violence that – at its worst – results in the complete annihilation of both of the feuding families (or, at least, all the males). The problem is that, just like with the low-level fictional feuds we have explored above, blood feuds rapidly take on a life of their own, with the triggering incident long forgotten. The first killing may have been so long ago that nobody alive today ever met the victim. Each modern-day killing is simply *gjakmarrja* for the last, which in turn was *gjakmarrja* for the one before that, and so on. Similarly in London, although the gang landscape is constantly shifting, rivalry between gangs from Brixton and Peckham dates back as far as the 1970s; perhaps even the 1950s.[22]

Can blood feuds be stopped? After the Second World War, Albania and other areas under the Kanun – including modern-day Montenegro, Kosovo, Macedonia and southern parts of Serbia, then all part of Yugoslavia – were ruled by Communist dictators Hoxha and Tito, under the direction of Moscow. The Albanian and Yugoslavian governments both officially banned and supressed the Kanun; but the blood feuds didn't end, they just lay dormant. Indeed, some experts argue that locals maintained the Kanun as a form of resistance against

---

* *Dr. Strangelove* is satire, but the Soviets really did build a device – Dead Hand – that automatically launches a retaliatory strike unless it is overridden by a human commander.

the 'occupiers'.[23] When the Communist governments of both countries finally collapsed in the early 1990s, the dormant feuds were reawakened.

The difficulty with blood feuds is that they can't be ended by somebody else; the feud is between the two families or the two street gangs. A powerful state can *suppress* the feud – it can put you in jail for exacting your *gjakmarrja* – but it can't *stop* it. It just doesn't have, and can never have, the authority to forgive your enemy on your behalf, or vice versa. No, the only way to end a blood feud is for the two sides to agree to do so. And if our exploration of the Feud masterplot has taught us anything, it's that feuds are ended only when the two sides unite to take on a bigger foe. This does happen occasionally with London street gangs. In *Street Casino*, Simon Harding discusses how, in South London, the GAS gang joined forces with Loughborough Bois and Murder Zone for protection against gangs from Peckham, Tulse Hill and Stockwell. But these alliances were only temporary; wholesale reconciliation in the face of a bigger foe happens only in the movies, right?

Yugoslavia (like, for example, the USA) was founded as a Federal Republic made up of individual states, in this case, six: Serbia, Croatia, Bosnia & Herzegovina, Macedonia, Slovenia and Montenegro. Kosovo, with its majority ethnic Albanian population, sat uneasily as an autonomous province within Serbia. As President of Yugoslavia, Tito maintained the peace with an iron fist, but tensions in Kosovo started to rise after his death in 1980, with many Serbs leaving the region. In 1990, the new President of Serbia, Slobodan Milošević, removed Kosovo's autonomous status, shut down Albanian-language newspapers and fired ethnic Albanian teachers, doctors and post-office workers.

Right up until his death in 1995, Anton Çetta lived and

breathed narrative. As head of the Department of Folklore in Kosovo's Institute of Albanian Studies, he published no fewer than sixteen books cataloguing the traditional stories of the region. Professor Çetta knew the power of narrative; its power to maintain feuds, yes, but also its power to heal them. Beginning in 1990, Çetta, together with around 500 student volunteers, went door to door in Kosovo, asking Albanians to forgive blood feuds. His method was described as a 'contagion of reconciliation'; Çetta would start out by looking for a relatively 'easy' case – for example, a feud that had gone cold while the threatened men waited out the *gjakmarrja* in the safety of their homes – broker a reconciliation, and then use it as an example at the next house. But the reconciliations were not just verbal agreements. The folklorist Çetta understood the need for a grand finale – the final scene of the movie where the former foes get together and have a great big party. Each reconciliation was a party with traditional music, dancing and food, attended not just by the newly reconciled families but friends, neighbours and local dignitaries. Most were relatively small affairs – perhaps thirty people or so – but the largest numbered well over 100,000 (grainy footage of this event survives on YouTube[24]). Altogether, around half a million people – that's a quarter of the entire population of Kosovo – attended a reconciliation. By 1992, somewhere between one and two thousand blood feuds had been forgiven; in 1993, not one single killing was recorded in Kosovo.[25]

How did Çetta and his students all but wipe out blood feuds in Kosovo? Narrative. From his years of immersing himself in stories, Çetta understood that **reconciliation and redemption** happens only when rivals set aside their differences to join forces against a deadly foe; in this case, Milošević and the Serbian army. As experts put it, 'it became an act of disloyalty against the resistance to pursue a blood feud against fellow Albanians in Kosovo';[26] families 'pardoned the blood . . . in the name of the youth, the people and the flag'.[27]

HAPPY ENDINGS

Overall, this chapter has been a fairly depressing one. Feud, we have learned, is etched into our DNA. From birth, we are on the lookout for those who are different from us, and respond positively to people who are mean to those who are 'not like us'. We are keen to divide ourselves into **evenly matched, mirror-imaged** opposing sides, even over something as trivial as which type of art we prefer, and to establish feuds that quickly take on **a life of their own**, persisting long after the original quarrel is forgotten. Although sporting feuds are usually good-natured, deadly feuds – such as those that play out on the streets of London or Kosovo – are depressingly common. Worse, although we are sometimes capable of setting aside our differences to take on a bigger enemy – like Milošević for Kosovans – the self-sustaining life-of-their-own nature of these feuds means that, in many cases – including, for example, London street gangs – there is no realistic prospect of them ending.

But it's not all doom and gloom. Although the Feud narrative is often the cause of death and destruction, it can also serve as a catalyst for human progress; particularly scientific progress. My own career is just one example.

So, as my mum likes to ask me from time to time, 'Ben, what is it that you actually do?' Well, my job title is 'Professor', which for most people suggests a kind of fancy teacher; one who lectures students in the finer points of philosophy, perhaps in an oak-panelled office with a decanter of sherry, probably wearing a tweed jacket with leather elbow patches (see, for example, the opening scenes of *Saltburn*). And, yes, teaching students is an important part of my job – I lead a course called 'Psychology in the Real World', based around popular psychology. But like most professors, assistant/associate professors, readers, senior lecturers and lecturers (essentially the same job at different levels of seniority), the majority of my job is research.

My particular research area – which you'd never guess from my popular-science writing – is children's language development. Most people, when I tell them this, assume that my research is something to do with helping children who struggle with language, or with reading, or with learning a foreign language. But it's none of those things. I'm fully aware of how ludicrous this sounds to an outsider, but all of my twenty-plus years of research, requiring several million pounds of funding, is geared around proposing and testing theories of how typically developing children learn their native language: that is, how a child growing up in the UK learns to speak English; in Israel, Hebrew; or in certain parts of Guatemala, K'iche' Mayan (all languages I've worked on[28]).

Where the rivalry comes in is that there are two main overarching theories of how children learn language, and the two sides have been engaged in constant battle since (at least!) the 1950s. As an example of the two theories, let's take a question my daughter came out with when she was two years old: '*What Mummy is doing?' Aww, how sweet. But wait! You'll notice there's an asterisk at the start of the sentence. This is standard linguistics jargon for 'WATCH OUT, THIS SENTENCE CONTAINS A GRAMMATICAL ERROR'. What my daughter meant to ask, of course, is 'What is Mummy doing?', but the middle two words were back to front, giving '*What Mummy is doing?' The question is why just about all English-speaking children make similar mistakes. And as with most things child-language related, the two rival theories offer very different explanations.

You've probably heard of Noam Chomsky – almost certainly the most famous academic alive; definitely the most highly cited. Well, the first view of how children learn language is very strongly associated with Chomsky; so much so that – at the risk of sidelining the many other researchers involved – I'll call this theory the Chomskyan view. Glossing over the details, what this theory claims is that children are *born with* a rule that takes a statement and turns it into a question by (Step 1) moving the question word to the front, then (Step 2) moving the 'auxiliary' (here, the word 'is'):

STATEMENT
Mummy is doing what →

STEP 1
What Mummy is doing? →

STEP 2
What is Mummy doing?

According to the Chomksyan account, children make mistakes like '*What Mummy is doing?' because, for reasons we needn't go into here, they get stuck after Step 1. But, so the theory goes, the fact that children always make *this particular mistake* (rather than, for example, asking '*Doing is what Mummy?') shows that they have the right rules for forming questions (in fact, that they were *born with* the right rules); they just get stuck sometimes.

That's the Chomskyan view. The opposing view, which I will

unimaginatively call the anti-Chomskyan view, says that language isn't about being born with rules, but about learning from what you hear. According to the anti-Chomskyan account, children make mistakes like '*What Mummy is doing?' because – again glossing over the details – they hear the combination *Mummy + is + doing* a lot ('**Mummy is doing** the delivery'; '**Mummy is doing** it, sweetie!'; 'You should really clear up your own toys, but **Mummy is doing** it'), and think that you can make a question just by slapping a question word on the start:

What + Mummy is doing →
*What Mummy is doing?

It sounds arcane, I know, but over the years, many experiments have been run to try to figure out which of the two explanations is right, or at least closest to the truth. I know because I've done many of them myself.[29] I won't bore you with the details but, very briefly, my experiments support the anti-Chomskyan view (at least, that's what I argue) because the more common the particular word combination in the language children hear around them (e.g. *Mummy + is + doing*), the more common the mistake (e.g. '*What Mummy is doing') in children's own speech.[30] This suggests, to my mind, that children make the mistake by learning from the language around them, rather than by getting stuck when applying an in-built rule. That is, I think the anti-Chomskyan view is the correct one. But many of my colleagues would disagree. Strongly.

And, let me tell you a little secret: the theoretical and experimental battles are the part I like. Developing a theory, figuring out what predictions it makes, designing an experiment to test those predictions, collecting the data, analysing the results statistically, writing up the findings persuasively, rebutting reviewers who disagree: *I love it*. Particularly when the results suggest that my theory is right and somebody else's is wrong. I try to make sure these battles are always professional rather than personal. Indeed, in the pre-Covid, pre-children days when I regularly attended overseas conferences, I would make a point of befriending and hanging out with 'the enemy'. But, nevertheless, these battles contain all the key ingredients of the Feud masterplot recipe.

First, the two sides are **evenly matched**: if you ask an academic which side is better supported by the evidence, they'll invariably say their own side, so that's not much help. But in terms of prestige and reputation, neither really has the edge. Since academics tend to hire others with similar viewpoints, universities – and even geographical regions – tend to be associated with one side or the other. In the US, very roughly speaking, the East Coast institutions like MIT, Harvard and the City University of New York tend to favour the Chomskyan approach, while Stanford, Berkeley and most of the University of California system tend to favour the opposite. In the UK, at least historically, University College London and its neighbours have tended to favour the Chomskyan approach, with Manchester, Liverpool, Nottingham and so on favouring the opposite. That is, it's not like one side has the Ivy League and Oxbridge while the other doesn't; it's pretty much an even split.

Second, the two sides are **mirror images** of one another. The anti-Chomskyan view gives, for my money, a convincing explanation of the kinds of mistakes that children make early in development, but – I will freely admit – says much less about how children eventually hit upon the adult system. The Chomskyan view, on the other hand, gives a detailed formal account of the adult system but, again for my money, does less well at explaining children's early errors. The two sides even have different *personalities*, like Cal and Aron in *East of Eden* or Maverick and Iceman in *Top Gun*. Taking a carefree, informal 'anything goes' approach to language is one of the biggest charges levelled against 'Maverick' anti-Chomskyan accounts by the formal and precise 'Iceman' Chomskyan accounts; and one that – I must admit – is not without merit.

Third, it's a bit embarrassing to admit, but – for me – the battle between the Chomskyan and anti-Chomskyan accounts certainly took on **a life of its own**, quite independent of the subject matter. What I lived for, professionally speaking, was devising, running and reporting experiments designed to test competing theories. The fact that those theories relate to child language development became almost incidental. Don't get me wrong, I find the topic completely fascinating. But I have little doubt that – if my professional life story had run a different course – I would have been just as satisfied developing and testing rival theories of, say, children's mathematical development, or how face recognition works in adults, or why we develop false memories.

But wait! What have we learned is the final key ingredient of the Feud recipe? That's right: in a final act of **reconciliation and redemption**, the foes eventually overcome their differences and work towards their common goal. And, do you know what? Maybe I'm getting soft in my old age, or maybe I've watched too many Feud movies, but I've certainly been moving in that direction. While I still design and run experiments constructed – to put it bluntly – to provide evidence for my view of how children learn language, I've started to focus less and less on explicitly arguing against the other side. In fact, my most recent project – comprising no fewer than five experiments and a computer model of the results – set aside this battle completely and instead contrasted two different accounts that sit under the same theoretical umbrella. After all, as we learned with Coke and Pepsi, since customers (in my case, fellow academics) find negative propaganda off-putting, you're probably better off just emphasizing the advantages of your own product and hoping that they like the taste.

But next time an academic tells you that their professional life story is defined by a Quest for knowledge (Chapter 2), Untangling some complex phenomenon (Chapter 3), or even slaying the Monster of some disease (Chapter 5), take it with a pinch of salt. Whatever they say, it's really a classic Feud.

7.

UNDERDOG

In a sunny spot stood an old country house, encircled by canals. Between the wall and the water's edge there grew huge burdock leaves. In this snug retirement, a duck was sitting on her nest to hatch her young.

At length, one egg cracked, and then another. 'Peep! Peep!' cried they, as each yolk became a live thing, and popped out its head. 'Well, how are you getting on?' inquired an old duck, who came to pay her a visit. 'This egg takes a deal of hatching,' answered the sitting duck. 'It won't break; but just look at the others, are they not the prettiest ducklings ever seen? They are the image of their father, who, by the bye, does not trouble himself to come and see me.' At length, the large egg cracked. 'Peep! peep!' squeaked the youngster, as he crept out. How big and ugly he was, to be sure!

The other ducks, after looking at them, only said aloud: 'Now, look! Here comes another set . . . and bless me! what a queer-looking chap one of the ducklings is, to be sure; we can't put up with him!' And one of the throng darted forward and bit him in the neck . . . 'He is too big and uncouth,' said the biting duck, 'and therefore he wants a thrashing.'

'Mamma has a sweet little family,' said the old duck; 'they are all pretty except one, who is rather ill-favoured. I wish mamma could polish him a bit.' 'I'm afraid that will be impossible, your grace,' said the mother of the ducklings. 'It's true, he is not pretty, but he has a very good disposition, and swims as well, or perhaps better than all the others put together.'

But the poor duckling . . . was bitten, pushed about, and made fame of, not only by the ducks, but by the hens . . . and a guinea-fowl . . . who puffed himself up like a vessel in full sail and flew at the duckling, and blustered till his head turned completely red.

Nor did matters mend the next day, or the following ones, but rather grew worse and worse. The poor duckling was hunted down by everybody. Even his sisters were so unkind to him that they were continually saying: 'I wish the cat would run away with you, you ugly creature!'. While his mother added: 'I wish you had never been born!'. And the ducks pecked at him, the hens struck him, and the girl who fed the poultry used to kick him.

So he ran away . . . And the duckling went forth, and swam on the water, and dived beneath its surface; but he was slighted by all the other animals, on account of his ugliness.

It would be too painful to tell of all the privations and misery that the duckling endured during the hard winter. He was lying in a marsh, amongst the reeds, when the sun again began to shine. The larks were singing, and the spring had set in in all its beauty . . . Oh, how beautiful everything looked in the first freshness of spring! Three magnificent white swans now emerged from the thicket before him; they flapped their wings and then swam lightly on the surface of the water. The duckling recognized the beautiful creatures and was impressed with feelings of melancholy . . .

'I will fly towards those royal birds, and they will strike me dead for daring to approach them, so ugly as I am! But it matters not. Better to be killed by them than to be pecked at by the ducks, beaten by the hens, pushed about by the girl that feeds the poultry, and to suffer want in the winter.' And he flew into the water . . . 'Do but kill me!' said the poor animal, as he bent his head down to the surface of the water, and awaited his doom. But what did he see in the clear stream? Why, his own image, which was no longer that of a heavy-looking dark grey bird, ugly and ill-favoured, but of a beautiful swan!

It matters not being born in a duck-yard, when one is hatched from a swan's egg!

He now rejoiced over all the misery and the straits he had endured, as it made him feel the full depth of the happiness that awaited him. And the large swans swam round him, and stroked him with their beaks.

Some little children now came into the garden, and threw bread-crumbs and corn into the water; and the youngest cried: 'There is a new one!' . . . And they clapped their hands and capered about . . . and more bread and cake was flung into the water; and all said: 'The new one is the prettiest. So young, and so lovely!'. And the elder swans bowed before him.

He was more than happy, yet none the prouder; for a good heart is never proud . . . He then flapped his wings, and raised his slender neck as he cried in the fulness of his heart: 'I never dreamed of such happiness while I was an ugly duckling'.

EXCERPT FROM 'THE UGLY DUCKLING' –
A TRANSLATION OF HANS CHRISTIAN ANDERSEN'S
'DEN GRIMME ÆLLING'[1]

The story you have just read is well over a hundred years old. But the Underdog masterplot recipe is as old as storytelling itself. The earliest written story that follows this template dates – give or take – to the year zero: the legend of Rhodopis, by the Greek geographer Strabo:

> When she was bathing, an eagle snatched one of Rhodopis's sandals from her maid and carried it to Memphis; and while the king was administering justice in the open air, the eagle,

when it arrived above his head, flung the sandal into his lap; and the king, stirred both by the beautiful shape of the sandal and by the strangeness of the occurrence, sent men in all directions into the country in quest of the woman who wore the sandal; and when she was found in the city of Naucratis, she was brought up to Memphis [and] became the wife of the king.

EXCERPT FROM STRABO'S *GEOGRAPHICA*, BOOK XVII, 33[2]

Rhodopis, of course, is the earliest known version of the Cinderella story. An Italian version known as *Cenerentola* was published in 1634. A French version, *Cendrillon*, followed in 1697, adding a pumpkin carriage, a fairy godmother and, no less improbably, changing the footwear to a glass slipper. A Grim(m) German version followed – *Aschenputtel*, 'Ash-fool' – in which the wicked stepsisters cut off their toes in order to make the slipper fit, before being blinded by doves (neither scene, needless to say, made Disney's cut). And those are just the European versions. *Ye Xian* (China), *Sumiyoshi* (Japan), *Tam and Can* (Vietnam) and *The Moon Brow* (Iran) are just a few of the others. In fact, as early as 1893, Marian Roalfe Cox, one of the first masterplot pioneers, was able to publish a volume documenting no fewer than 345 variants of the Cinderella story.[3]

We have been telling the same story ever since, highbrow and lowbrow, in oral folklore, plays, novels, movies and even video games: *Aladdin, King Arthur, Dick Whittington, Jane Eyre, David Copperfield, Babe, Pygmalion, Charlie and the Chocolate Factory, Joseph and the Amazing Technicolor Dreamcoat, Slumdog Millionaire, Grand Theft Auto IV* . . . all follow this most compelling of masterplot recipes.

THE MASTERPLOT RECIPE

The story that I chose to introduce the Underdog masterplot – Hans Christian Andersen's 'The Ugly Duckling' (first published in Danish as 'Den Grimme Ælling' in 1843) – is a particularly clear illustration of this recipe: while human Underdog stories distract us with all kinds of irrelevant seasonings like pumpkin carriages and glass slippers, Andersen's follows the Underdog recipe in its most concentrated form: all killer, no filler.

The most important ingredient of the Underdog recipe, which must be added at the very climax of the story, is **destiny realized** (or, to use a more modern phrase, 'finding yourself'). At the start of the story, our would-be hero has some ineffable essence of greatness – nobility, kind-heartedness, incorruptibility, leadership skills or pure rizz – but their light is hidden under a bushel. At the climax of the story, this essence of greatness becomes obvious to both the hero and everyone around them. 'The Ugly Duckling' is a particularly clear example, because the essence of greatness is straight-up genetics: literally being a swan. Interestingly, most experts think 'The Ugly Duckling' is either about Andersen himself – who had not just literary talent but possible royal blood – or his friend, opera singer Jenny Lind (who, 175 years later, was portrayed by Rebecca Ferguson in *The Greatest Showman*). For Cinderella, this essence of greatness is her 'exceptionally sweet and gentle nature. She got this from her mother, who had been the nicest person in the world'.[4] For *Slumdog*'s Jamal, it's his humility, his determination, his resourcefulness. For *Charlie and the Chocolate Factory*'s Charlie Bucket it's his generosity and selflessness (as shown, for example, when he shares his precious birthday chocolate – the one bar he gets each year – with his extended family).

UNDERDOG

It's worth noting at this point that the Underdog recipe is similar to the Monster recipe, though there are two key differences. The first is that the underdog doesn't do battle with an all-powerful monster, not even a metaphorical one. Sure, they defy the odds to achieve an improbable success, but they don't have to kill anyone; there's no monster threatening the kingdom. As a result, there's no sense – as there is in Monster – that our underdog hero is acting on behalf of their whole group or tribe; their eventual triumph is a personal one. The second key difference between the two masterplots relates to this crucial ingredient of **destiny realized**. Yes, the protagonist of a Monster story is an unlikely hero, but there is no ineffable essence of greatness; no inevitable destiny realized. If 'The Ugly Duckling' were a Monster story, as opposed to an Underdog one, the duckling wouldn't be a swan; just a normal duckling who took it upon himself to – let's say – defend the duck pond from the neighbourhood cat.

So important is this key ingredient of **destiny realized** that the other key ingredients – important though they are – serve mainly to bring out the flavour of this primary ingredient. Take, for example, the most obvious key ingredient: **humble beginnings**.*
Why exactly must the story start with our hero in poverty (Charlie Bucket, Jamal Malik, *Grand Theft Auto IV*'s Niko Bellic), at the bottom of the heap, bullied by those around them (Cinderella, the Ugly Duckling)? The reason is precisely to ensure that this essence of greatness remains well hidden until it is dramatically revealed at the climax of the story. A story in which a king-in-waiting enjoys all the comforts and trappings of royalty before smoothly transitioning to power when his father dies is no story at all.

---

* Again, Hans Christian Andersen's own life story fits 'The Ugly Duckling' to a tee; although he claimed royal blood, his father was a poor shoemaker, and his mother a washerwoman.

The same is true for another key ingredient: **meeting the hero as a young child** (Cinderella, Charlie Bucket, Jamal Malik), or even (as in 'The Ugly Duckling') as a newborn. Again, this ingredient serves to bring out the flavour of **destiny realized**: it's important for us to meet the hero as a child, so we can see that this ineffable essence of greatness that eventually comes out *was there all along* ('She got this from her mother, who had been the nicest person in the world'). Again, the Ugly Duckling (described as having 'a good heart', 'a very good disposition') is a particularly clear example because it is so literal: we meet him as an egg, but an egg that is far too big to plausibly be a duck's egg. His essence of swanness is there all along. It's just the same for human heroes of Underdog stories. Cinderella, Charlie and Jamal didn't *learn* their sweetness, selflessness and humility any more than the Ugly Duckling learned his swanness; they were born with it. Occasionally, we first meet the hero as an adult, but they are still in some important way childlike; *Grand Theft Auto IV*'s Niko Bellic is an adult, but he arrives in Liberty City fresh off the boat, as much an ingénu as if he had literally just hatched out of an egg.

Another key ingredient that serves mainly to bring out the recipe's **destiny realized** flavour is **the ugly sisters**. In both *Cinderella* and 'The Ugly Duckling', they are literally sisters (with, in the latter case, chickens, the farm girl and even the poor little thing's own mother also playing the ugly sister role). Andersen himself was apparently mocked by his classmates for his 'big nose . . . big feet . . . beautiful singing voice and a passion for the theatre'.[5] In *Charlie and the Chocolate Factory*, the ugly sisters are the other children who join him on the factory tour, who are all greedy and selfish. In *David Copperfield*, they are David's cruel stepfather and step-aunt (named, in typical Dickensian fashion, the Murdstones). In *Slumdog*, it's Maman, the exploitative leader of a gang of street children. The role of the **ugly sisters** is to highlight the hero's ineffable positive qualities, which are brought out at the climax where we see **destiny realized**. The ugly sisters do this not only by being a foil to the hero – selfish when our hero is selfless, mean when our hero is generous and so on – but also, and perhaps most crucially, by being blinkered to our hero's ineffable essence of greatness. So cruel and vain are the sisters, it doesn't occur to them that Cinderella could be not just sweet, but beautiful. So greedy and selfish are the other children touring the chocolate factory, it doesn't occur to them that Charlie could be modest and selfless. So ducklike are the ugly duckling's siblings that it doesn't occur to them that he could be a swan.

The final ingredient of the Underdog masterplot recipe, which must be added more or less straight after **destiny realized**, is **the reward**. Again, this ingredient's main function is to bring out the flavour of **destiny realized**. It's not enough for Cinderella's sweet and gentle nature, Jamal's humility, determination and resourcefulness, Charlie's generosity and selflessness or the ugly duckling's swanness to just become widely noted. These qualities must be rewarded with the prince's hand in marriage, *Who Wants to be a Millionaire*'s jackpot, Charlie's very own chocolate factory or copious amounts of bread and cake (well, what else does a swan like?). Only then can the humble underdog be said to have truly realized their destiny.

STRANGER THAN FICTION

In 1485, King Richard III (you know, the baddie from the Shakespeare play) was defeated and killed at the Battle of Bosworth, marking the end of both the Plantagenet dynasty and the Middle Ages. Richard was slung naked over the back of a horse and unceremoniously buried in a provincial churchyard. When the Tudor king Henry VIII (you know, the one with all the wives) ordered the dissolution of the monasteries fifty years later, Richard's remains were thrown into the River Soar. Or so it was widely believed.

Saturday night. BBC One. Prime time. A man stripped down to his underwear sweats under the studio lights, while his colleagues snigger and mock. A scene from Netflix's dystopian near-future series *Black Mirror*? Five hundred and thirty-one years and one week after the Battle of Bosworth, England footballer turned TV host Gary Lineker made good on his promise to present *Match of the Day* in his underpants if his team, Leicester City, won the Premier League title. (Sort of. Most commentators – and there were many – took the view that they were actually shorts.)

Richard III's remains were discovered under a Leicester car park in 2012. On 26 March 2015, he was finally given a decent burial at the city's cathedral. The previous Saturday, Leicester City had lost to Tottenham Hotspur in the Premier League, leaving them bottom of the league table with a mere nineteen points. This is where we **meet our hero as a young child**. Despite their lowly league position, the raw material – the ineffable essence of greatness – was there all along. Jamie Vardy, the leading goal scorer, Riyad Mahrez, the Premier League Player of the Year in the title-winning season, and Kasper Schmeichel, their heroic goalkeeper, were already there. But this greatness remained very much hidden. In fact, with just nine matches left to play, no team in Leicester's position had ever escaped relegation. But with the city's most famous son now finally resting in peace, escape they did, and comfortably: Leicester won seven of those nine games, losing only to eventual title winners Chelsea.

UNDERDOG

When the next season started, Leicester simply carried on where they left off, winning twenty-three of their thirty-eight matches and losing just three. They won the Premier League title not by scraping over the line on the last day of the season but by a clear ten points, earning as their **reward** immortality as one of only a handful of clubs to win the Premier League (as well as a cool £100 million).[6] Some, including the city's mayor, put Leicester's turnaround to Richard III. Others pointed to the influence of the club's owner Vichai Srivaddhanaprabha, whose duty-free business was prominently advertised across the players' chests. Its name? King Power – you can't make it up.

Leicester City's Premier League win of 2015–16 is, most commentators would agree, the greatest underdog triumph in the modern era of team sports. It is difficult to convey, especially to readers with no particular interest in football, just how unlikely it was, at the start of the season, that Leicester City would end up as champions. The often-quoted statistic that bookmakers offered odds of 5,000:1 doesn't quite capture their **humble beginnings**. A bookie's job is to offer odds that persuade punters to part with their cash. For extremely improbable events, this is less a matter of calculating precise probabilities than of plucking a nice juicy round number out of thin air. 5,000:1 is a 'go-to': the odds typically offered for everything from catching the Loch Ness Monster to uncovering an octogenarian Elvis; from Barack Obama playing cricket for England to Kim Kardashian being elected the next President of the United States.[7] More informative are the odds that were being offered on Leicester's relegation. They were not quite the relegation *favourites* but, at 3:1, the consensus was clear: Leicester were almost certain to finish closer to the bottom of the table than the top.

Another statistic that comes close to capturing Leicester's **humble beginnings** is the cost of their squad. Overall, it was around £54 million, though in fact some of the most expensive individuals were bit-part players. The two biggest stars, Vardy and Mahrez, cost just £1 million and £0.35 million respectively; less than the club spent on cardboard clappers, dished out free to fans at home games to boost the atmosphere. In contrast, **ugly sisters** Chelsea, who won the league the season before, had spent more than the entire cost of Leicester's squad on a single player, Diego Costa (likewise for fellow **ugly sisters** Manchester City and Kevin De Bruyne). But, again, the figures don't quite tell the whole underdog story. For example, the majority of Barcelona's 2009 European Champions League winning team cost them little or nothing – home-grown products of the famous 'La Masia' academy – but they were certainly no underdogs.

What about the wage bill? Now we're talking. In *Soccernomics*, Simon Kuper and Stefan Szymanski demonstrate that a club's wage bill is by far the best predictor of its success. For all the talk of managers and tactics, the correlation between wage bill and league position is almost perfect: around 90 per cent. In the eighteen seasons prior to Leicester's title win, every single league-winning team had ranked in the top three on total wage spend. In that season, Leicester (£80 million) ranked sixteenth out of twenty (Manchester United were top with £232 million).

Yet even their low wage bill doesn't quite tell how remarkable Leicester's story was. To an outsider, it might seem that Leicester were simply very lucky, or very shrewd, with their signings. After all, Riyad Mahrez soon moved on to Manchester City for a fee of £60 million, where his personal wage bill went up to around £10 million a year. To an outsider, it might seem that Leicester somehow managed to snap up all of the best players around. But that's not it. With the exception of Mahrez (and possibly N'Golo Kanté), you would struggle to name a single Leicester player who would have been regarded as the top Premier League player in his position.

Leicester City's Premier League win is the greatest ever Underdog story in team sport because, like Cinderella, they stayed true to their heart. They did it their way. Their ineffable essence of greatness, which eventually saw their **destiny realized**, was a style of play that flew in the face of convention.[8] Every other team who won the Premier League in that era played, by comparison, a game of slow and intricate build-up, with an average of around 4.5 passes per spell of possession, advancing towards the opponents' goal at a speed of just over 1 metre per second. On both of these metrics, Leicester were huge outliers amongst Premier League champions with an average of under 2.5 passes per spell of possession, advancing towards the opponents' goal at a lightning quick speed of around 2 metres per second. Leicester's speed and directness, then, turned out to be their ineffable essence of greatness.

But there is another crucial ingredient in Leicester City's success: the Underdog masterplot itself.

THE SCIENCE BEHIND THE STORY

At first, everyone thought Leicester's league position was an anomaly. But gradually, as the season wore on, and they never dropped below second, the Underdog masterplot took over. Yes, they had started the season as huge underdogs but, yes, this was really going to happen; a team really could go from outsiders to favourites over the course of a single season.

Self-fulfilling-prophecy effects (or 'expectancy effects') crop up all over the place in academic psychology, as well as in entirely unrelated areas of study. The best well known are placebo effects. For example, if you think that an antidepressant drug will make you feel better, then – on average – it probably will, even if the 'drug' turns out to be nothing more than a sugar pill. The only thing that made you feel better was the expectation that you'd feel better: a self-fulfilling prophecy. And it's not as if the placebo tricks you into somehow thinking you feel better when you actually don't. A 2022 meta-analysis (a study which combines data from all previous studies on a particular topic) found that antidepressant placebos act on many of the same areas of the brain as actual antidepressant drugs.[9] Placebo effects are found in sport too, with a 2019 systematic review (similar to a meta-analysis) finding that placebo nutritional supplements actually helped athletes to run or cycle faster, or to lift heavier weights.[10]

One area in which self-fulfilling-prophecy effects are particularly commonly studied is education. For example, a German study published in 2020 asked class teachers to rate each of their students in reading and maths.[11] The researchers also gave the students standardized reading and maths tests, to establish how good they *actually* were, as opposed to how good teachers thought they were. For some students, the teachers were spot on. But some students were actually a bit better than their teachers thought they were; some were actually a bit worse. At the end of the school year, the students completed the standardized reading and maths tests again. The results were astonishing: the students whose ability had been overestimated by their teachers did better than they ought to have done (given their earlier test scores), while the students whose ability had been underestimated did worse. Of course, the students' actual abilities and efforts mattered too, but the teacher's expectations were a classic self-fulfilling prophecy: if your teacher thinks you're actually better at maths than you are, then you rise to meet their expectations.

Something very similar happened in the 2015–16 title race, as the Underdog narrative took hold not only in the minds of the manager and players at Leicester City, but also in the minds of their opponents. This was exemplified by Leicester's Valentine's Day defeat at Arsenal (one of just three all season). From the home dressing room, selfies and videos surfaced of Arsenal players celebrating as though they had just beaten not relegation favourites, but title contenders. *Hold on*, thought the Leicester players, just like the German primary school students, *they think we're actually better than we are. But maybe they're right. Maybe we really are Premier League champions in waiting.*

Perhaps most crucially of all, this Underdog narrative took hold in a third group: referees. Now, don't get me wrong, I am not saying for one moment that Leicester City didn't deserve to win the Premier League. On the contrary, for beating the glamour clubs at their own game on a shoestring budget, Leicester more than deserve their triumph. But, beyond the 5,000:1 odds and the £54 million squad cost, there is another statistic that is less widely known. In 2015–16, Leicester equalled the record for the most penalties awarded in a Premier League season: thirteen. This is unusually high. For example, Leicester's main title rivals, Tottenham and Arsenal, were awarded just five and two. It is true that Leicester had one player who – to put it politely – had a particular knack for winning penalties (opposition supporters generally put it less politely). But referees sold on the Underdog story can be forgiven for going easy on Jamie Vardy, a player Leicester plucked from the fifth tier of English football. (The one referee who did book Vardy for diving – and it was his second yellow of the game – promptly gave Leicester a highly questionable penalty in the fifth minute of injury time, securing them a valuable draw in the crucial final stages of the season.)

It's not that the penalties themselves made the difference: three were missed, two were awarded after Leicester were already champions and one was in a game they lost anyway. It's what they represent: referees who – while not *consciously* trying to aid Leicester's Underdog story – would have to be damn sure before dealing it a possibly fatal blow. In every match, a referee makes dozens of close calls that could go either way, not just 'Is it a penalty?' but 'Is he offside?', 'Is that a free kick?' etc. Given the razor-thin margins of professional football, all it takes – even for just a fraction of those close decisions – is for a referee to err, subconsciously, on the side of caution; to avoid upsetting the applecart of history trundling inevitably along the path from rags to riches.

The self-fulfilling-prophecy effect is unlikely to be the only psychological effect at play here. A related effect that probably helped Leicester City realize their destiny is what psychologists call the status-quo bias. A number of experimental studies have shown that when asked to choose between a number of different hypothetical options – for example, for electricity-supply contracts[12] – more than half of participants choose the option designated by the experimenter as the 'status quo', even when the set-up is entirely hypothetical, meaning that no greater effort is required to 'twist' than 'stick'. Thus, once 'Leicester are really going to win this' had become the status quo for the season, it became psychologically difficult for referees to undermine it even though, of course, they had no real skin in the game.

Another relevant, and related, bias is risk aversion. Simply put, when faced with uncertainty, we generally choose the option that seems less risky. Suppose a referee is 50/50 as to whether a foul on Leicester forward Jamie Vardy warrants a penalty. In a season where neither Leicester nor their opponents are title contenders, neither option is particularly risky; the referee can simply go with his gut instinct. However, in a season increasingly defined by the Underdog masterplot – in the minds of Leicester players, their opponents, football fans, casual observers, pundits, newspaper editors and even the referee himself – declining to award the penalty incurs a huge risk (that of becoming the most hated man in football), and for what?

UNDER THE INFLUENCE

Leicester City's improbable Premier League win shows what can happen when an Underdog narrative unconsciously takes hold. But what about when someone quite deliberately uses the Underdog masterplot to consciously shape human behaviour?

> Citrain IPA began when millionaire investment banker Simon Cooper decided to muscle in on an already-overcrowded marketplace. Cooper was no fan of pale ale, or, for that matter, of alcoholic beverages in general. But he noticed that while customers were increasingly willing to pay premium prices for IPAs, no brand had yet established dominance in any of the world's major markets. Calling on all of his contacts, and calling in all of his favours, Cooper was able to raise $200 million for the launch. Initial sales, though a little sluggish, were well within the lower range of projections, and Citrain remains on course to turn its first profit in 2028.

Have you ever read an origin story like this? While it might be a pretty accurate one for most brands launched in recent years, this is not one that you'll find on the bottle. If Citrain IPA were real – if you haven't guessed, I just invented it, using an online brand-name generator[13] – its spiel would be more along these lines:

THE STORIES OF YOUR LIFE

Citrain IPA began life in the 2008 financial crisis. Unable to find a job after leaving university (**meeting the hero as a [metaphorical] child**), Simon Cooper decided to indulge his lifelong passion for real ale (**his destiny to be realized**), and started brewing small batches of IPA in his mother's kitchen (**humble beginnings**). Word quickly got around about the unique taste of Citrain, and it wasn't long before MegaBrewery International (**ugly sisters**) got in touch with an offer for the company. Unhappy with MBI's plan to brew Citrain in its Manchester mega-plant, Simon took the decision to go it alone, and got his **reward** in 2024 when Citrain was named Britain's Best Beer by the National Real Ale Society (**destiny realized**).

That's right, when companies are launching a new product and need a story, the Underdog masterplot recipe is more often than not their go-to. If you think the fictitious example above sounds far-fetched, compare it to this entirely genuine one below:

Legend has it that Frankie Giuliani was 10 years old (**meeting the hero as a child**) when, with his Mamma and Poppa, he left Sicily and landed at Ellis Island, New York in 1924. They moved in with relatives (**humble beginnings**) in Little Italy. From the home country Poppa brought a little money (more **humble beginnings**) and a lot of ambition (**his destiny to be realized**). It was no surprise, then, when the family opened a restaurant within a year. It was Mamma's home-style cooking and special recipe sauces that kept the regulars comin' back for more (**destiny realized and reward**). In 1953 Poppa retired and Frankie took over the business with his ol' school pal Benny (**destiny realized**).[14]

'Legend has it', indeed. In fact, the original Frankie & Benny's was opened by City Centre Restaurants PLC (now The Restaurant

Group PLC) in 1995, in the UK, just a stone's throw from Leicester's most famous car park. There has never been a Frankie & Benny's in New York, or indeed anywhere outside of the UK. And while much of his popular back story is dubious, at least Richard III was a genuine historical figure, in the sense that he – you know – actually existed. I have no idea exactly whose face is plastered all over the walls and the menus of Frankie & Benny's restaurants, but it is certainly not that of Frankie Giuliani (come on, were you even trying?) or his ol' school pal Benny ('Sorry, chaps, we're fresh out of Italian surnames').

For a more authentic story of an immigrant family made good, consider the Wozniaks. Steve Wozniak, with his ol' school pal Steve Jobs, created what is now the biggest company in the world, Apple, in Jobs's mum's garage at 2066 Crist Drive, Los Altos, California – now officially designated as a historical site by the city council.[15] Here's how Wozniak remembered it in 2016:

> We never once discussed a product in the garage, never conceived of a product, never talked about features of a product in a garage – we did in a lot of other places – but people thought we had a garage with people sitting around in it. No.[16]

Oh. Why not? Well, Jobs's garage – like most garages – came attached to something called a 'house', a structure that superficially has much in common with a garage but includes amenities such as chairs, a bathroom, a kitchen and a telephone line to the outside world. Wozniak's family also had one of these so-called 'houses' which, for some reason nobody could quite put their finger on, felt better suited to product development meetings than Steve's mum's garage, particularly in the sweltering California summers.

THE STORIES OF YOUR LIFE

Where did the myth of the garage come from? It turns out that, forty years ago, the two Steves had already cottoned on not only to the importance of the **humble beginnings** ingredient of the Underdog masterplot – which in their case was largely true – but to coming up with a way to embody that story in an image that was visual, tactile, even olfactory: the suburban California garage in the sunset, looking like a shot from *ET*, the heat coming off the garage door, the smell of engine oil on the concrete. Here's Wozniak again:

> The garage represents us better than anything else . . . We would drive the finished products to the garage, make 'em work, and then we'd drive 'em down to the store that paid us cash.[17]

It's Apple's entire operation in a single metaphor. Whereas most companies drive the finished product straight to the store, Apple takes a detour via the garage to scuff up the brushed aluminium with a pad of outsider chic. It was the same story with Apple's 1997 'Think Different' campaign, whose slogan was a direct retort to IBM's 'Think IBM' ThinkPad campaign:

> Here's to the crazy ones.
> The misfits.
> The rebels.
> The troublemakers.
> The round pegs in the square holes.

Contrary to popular belief Steve Jobs didn't write a word of the commercial. In fact, at least according to the creative director of the ad team, Jobs initially called it 'advertising agency shit'.[18] But, to his credit, he eventually realized that it was the perfect next scene in the movie that began with the shot of his mum's garage door. The campaign paved the way for the brightly coloured iMac G3, which looked like no computer before or since, and which signalled not only the rebirth of Apple but also the beginning of the end of the record store: 'Rip MP3s. Burn CDs. Fast.' urged the 'Flower Power' commercial, over a skuzzed-up grunge cover of The Smiths' 'How Soon Is Now?'

The '"I'm a Mac", "I'm a PC"' adverts of the mid-2000s were less a brand-new campaign than another chapter in the same old story: the Mac is the Underdog outsider; the PC is the ugly sisters, The Man. Around eighty different adverts were made, but they're all essentially the same (I'm paraphrasing here):

> PC: I have to say that iTunes is pretty good.
>
> Mac: Thanks! The rest of iLife is just as great and comes on every Mac!
>
> PC: [sheepishly] I come with lots of cool apps too: calculator . . . clock . . .
>
> Mac: I enjoy doing fun activities like making podcasts and movies.
>
> PC: [defensively] Hey, I can do fun activities, too, like timesheets, spreadsheets and pie charts.

In reality, Bill Gates of Microsoft, at whom these clever ads took aim, had a very similar back story to Jobs: both were high-school bedroom tinkerers who dropped out of college to pursue their passion for personal computers. The difference is that even as Newton-defying Apple rose to the very top of the corporate tree, it never gave up – or even eased up on – its Underdog shtick. One of its more recent campaigns, Behind the Mac, ostensibly flagged up the laptop's unique potential for making music. At one level, it's pure bluster: a PC can run the industry-standard Pro Tools software just as well as a Mac, and you'll struggle to find a professional who uses Apple's Logic Pro, let alone its Garage Band. Yet on another level it's spot on: you *will* struggle to find a laptop musician who uses a PC because, although Apple is worth more than pretty much all PC manufacturers put together, it's somehow still the cool Underdog.

What Apple's story tells us, especially in comparison to Microsoft's, is that, in order to stick, a narrative doesn't have to be unique, true or even *believed* to be true. It just needs to resonate. People who buy a Mac or an iPad don't genuinely believe they're somehow sticking it to The Man. They know full well that Apple and Microsoft are pretty much neck and neck in terms of their valuation (if anything, Apple is currently slightly ahead).[19] But they know that Apple has this Underdog narrative, and that everyone else knows this, too; it's a shared narrative that – like Leicester City's Underdog story of 2015–16 – you can't just step outside of. I once asked an American professor friend why, in the US, even impoverished students overwhelmingly buy Macs over much cheaper PCs that are easily up to scratch for college work. 'In the US,' he told me, 'PCs are for Republicans.' What Apple's story tells us, then, is that masterplots – even ones that nobody really believes – can be cleverly utilized, whether to sell computers or something much bigger, even by Republicans . . .

PLOT TWISTED

For our next Underdog story, we must now turn to a certain billionaire businessman; a self-made man whose knowledge of 'the art of the deal' brought him almost unimaginable wealth. Trump was certainly shrewd. His masterstroke was to see the opportunities made available by government-backed housebuilding schemes. His modus operandi was to take out housebuilding loans from the Federal Housing Administration, on much more favourable terms than were available on the open market, and then build the agreed development at a significantly lower cost. Trump didn't simply pocket the difference, which of course would have been illegal, but invested the money elsewhere. Unethical? Well, that depends on your ethics. Illegal? No. A Senate Committee on Banking and Currency tried to nail him for allegedly taking out a loan some $2 million dollars larger than the actual construction costs on an apartment complex, but found that no rules had been broken. The transcripts make for amusing reading, filled as they are with characteristic Trump swagger and bluster.[20] Trump, of course, arrives three hours late to the hearing . . .

Mr. SIMON (general counsel, Federal Housing Administration Investigation): Mr. Trump, what was the cost of the construction of section 1?

Mr. TRUMP: You want the actual cost with interest on the advances?

The CHAIRMAN (Senator Homer E. Capehart): Is there any cost other than the actual costs? You have just one set of costs, don't you?

Mr. TRUMP: There is a difference, Senator.

Mr SIMON: We want all the money you paid out to anybody to construct section 1.

The CHAIRMAN: When all the bills are paid, what was the total?

MR TRUMP:	How about real-estate taxes on the land during construction? Interest on buildings?
The CHAIRMAN:	You have the figures. Just tell us the total cost after it was all paid.
Mr. TRUMP:	We have here Beach Haven Apartments 1, schedule of construction costs, and it totals $4,015,783.
Mr. SIMON:	You don't want to be misunderstood, testifying under oath, that you paid that money out, do you, Mr. Trump?
Mr. TRUMP:	No, I will explain this to you, Mr. Simon.
Mr. SIMON:	What were your costs?
The CHAIRMAN:	Your actual costs we want. Not the fees that you didn't pay, such as these architects' fees and builders' fees. What did it actually cost you in dollars and cents, please? Give us that, will you please?
Mr. SIMON:	Did it cost you $3,627,332?
Mr. TRUMP:	I would say roughly without figuring builders' fees, which we took ourselves.
Mr. SIMON:	Did you pay any builders' fee?
Mr. TRUMP:	We absorbed – we did the work you would ordinarily pay a builders' fee for and we are entitled to the builders' fee because the project was built. We performed the service.
Mr. SIMON:	When you mow your own lawn, does anybody pay you a fee for it. . . . You own the building, is that right?

Mr. TRUMP: If a tailor has one of his men make a suit of clothes, that suit will cost X dollars. If the boss tailor makes a suit of clothes, he can't sell that suit cheaper. That suit is worth just as much as though he paid a man to make the suit . . .

There are pages and pages of this stuff, but eventually they get it out of him.

What did Trump do with his vast wealth? He liked to spoil his children, to the tune of around a billion dollars, split between them.[21] One of them, Donald, who would later go on to become President of the United States, received around $400 million. What, you thought *Donald* was the self-made man? No, that was his father, Fred. Although Donald did make some smart bets – in particular, getting into Manhattan property when everyone else was getting out – he was given more than a little help from the Bank of Mum and Dad, not to mention a billion-dollar bailout from seventy banks in 1990.[22] If you or I were in the hole to the tune of a billion dollars, we'd find it difficult to somehow convince the world that we were business geniuses.

Not Donald Trump. For, whatever his shortcomings, Trump Jr understands the power of narrative. He realized that simply *playing the role* of an Underdog, self-made businessman on a popular TV show, *The Apprentice*, would be enough to convince the public that he was the real deal. He repeated the trick in his 2016 election campaign by framing himself as the plucky – if, by now, rich – Underdog against 'establishment' Hillary Clinton.

Does any of this matter? Yes. A poll conducted in the wake of his 2016 election victory found that the mistaken belief that Donald Trump was *not* born into wealth was worth about five percentage points in his approval ratings (that **humble beginnings** ingredient again).²³ A follow-up study a year later found that, when given a brief summary of his real back story, even Republicans showed a 9–10 per cent drop in their ratings of Donald's empathy and business competence.

Elections can turn on such margins: remember that in 2016, Trump actually *lost* the popular vote by a bigger margin than any other election winner. The lesson is that playing along with an Underdog narrative that we know to be false isn't just a bit of harmless fun. If we allow ourselves to become duped by such a narrative, even knowingly, we risk feeding into it. By and large, Democrat politicians and supporters were happy to laugh along: 'Donald *Trump*. Are they *serious*? Come *on*.' These kinds of jibes fed Trump's Underdog narrative, allowing him to play to his base: the little guy who is overlooked by the governing establishment.

At the time of writing, Trump had just sealed the Republican nomination for 2024. By the time this book comes out, the election will be just weeks away. If he wins, Trump has promised to be a dictator on Day 1,²⁴ and given the events of 6 January 2021, many sober and serious commentators are saying that a second Trump term could see the end of US democracy. If that happens, you can blame the Underdog narrative that Trump has already used to make a name for himself on *The Apprentice* and to beat Hillary Clinton, and that he will no doubt seek to use again to beat the sitting president, Joe Biden. Don't say I didn't warn you.

HAPPY ENDINGS

A story that follows the Underdog recipe climaxes with our hero's **destiny realized**: the ineffable essence of greatness that was clear when we **met the hero as a young child** – nobility, kind-heartedness, incorruptibility – is finally brought out into the open for all to see. It's all a far cry from our hero's **humble beginnings**, when they started out at the bottom of the heap, shunned and put upon by a gaggle of **ugly sisters**. Now, though, the hero finally gets their reward: the keys to the kingdom, or to the heart of the beautiful prince(ss).

Because we all root for the Underdog, it's a masterplot recipe that is ripe for exploitation, by those selling everything from IPA to chicken wings to Apple Macs to right-wing politics. But when a genuine Underdog narrative takes hold, the results are heart-warming, as when Leicester City defied the odds to win the Premier League title, or as in the story of Jyoti Mishra.

When I was first kicking around ideas for this book, I went to visit Jyoti at his house in Derby (as it happens, just twenty-five miles down the road from Leicester). As with all Underdog stories, he started by telling me about his childhood. When he was six, Jyoti's mum organized a birthday party for his classmates. She didn't make any Indian snacks, because she wanted the family to seem 'normal', to 'fit in'. Crisps, chocolate fingers, party rings, sandwiches made with white bread. Nobody turned up. Shunned by his classmates and overshadowed by his sister – who followed their parents into medicine – Jyoti was the very definition of an Underdog. 'I've always been an outsider, and always not belonged, in anything that I've done,' he told me. 'So you naturally gravitate towards underdog kind of stories. It's why I love comics. My favourite thing when I was a kid was Spider-Man, because he wasn't Tony Stark [Marvel's Iron Man – a billionaire who inherits wealth from his father]. He made mistakes, he did the wrong thing, he fell in love and got worried about missing school and stuff like that.'

But the ineffable essence of greatness was there: a passion for computers ('a whole field of people who were geeks at school winning out in the end') and music. If only there were some way to merge the two. By the mid-1990s, there was. Just about. The technology was primitive by today's standards, but Jyoti scraped together just about enough money to get a Tascam 688 four-track tape recorder and an E-Mu Emax II sampling keyboard. He used the Emax to sample a trumpet part from Lew Stone and the Monseigneur Band's 1932 song 'My Woman' (as Dua Lipa would do for her 2020 hit 'Love Again'), and a loop from a breakbeat samples CD. Over this unlikely and ramshackle backing track, Jyoti sang about being an outsider; about being a Marxist in a capitalist world, a straight guy in love with a lesbian. Both his act, White Town ('No Black or Asian people have ever asked me what it meant'), and his album, *Women in Technology*, were deliberate references to Jyoti's outsider, Underdog status.

You know how this ends. Jyoti's single went to number one in the UK (and Spain, Israel and Iceland), and even number five in the US. His reward? The keys to the kingdom. The house in whose kitchen Jyoti made me eggy bread had been paid for entirely with his 'Your Woman' royalties.[25]

Jyoti's story is a powerful illustration of the power of the Underdog masterplot recipe as a catalyst for human progress. It's not just that he fits the Underdog story. Like Leicester City, Jyoti consciously used the Underdog recipe to inspire himself: he sang about being an underdog.

But what about the rest of us? Those of us who aren't necessarily outsiders or underdogs ourselves. What can the masterplot recipe offer us?

Professor Melvin Lerner sums up the answer in the title of his 1980 book *The Belief in a Just World: A Fundamental Delusion*, a classic of social psychology.[26] Twenty-five years' worth of studies across a range of different cultures point to a single conclusion: despite all of the available evidence to the contrary, the vast majority of us continue to believe that good things tend to happen to good people and bad things to bad people. Of course, our rational selves know full well that this 'just-world' belief is nonsense. But our subconscious attitudes and beliefs show that we just can't shake it. For example, in a typical study,[27] participants were shown a news report of a mugging and then told either that the muggers had been caught, or that they had not been caught and almost certainly never would be. The researchers then had participants complete various tasks and questionnaires designed to tap into their subconscious attitudes.

Remarkably, telling people that the perpetrator was still at large made them rate the victim as more careless and irresponsible. After all – goes the just-world belief – if the victim was *completely* blameless, then the universe would have conspired to ensure that their attacker was caught. Similarly, telling people that the perpetrator was still at large made them rate themselves as less likely to be the victims of such a crime. Again, according to the just-world belief, victims are at least partly responsible for their own misfortune. So – the thinking goes – as long as I keep being careful and keep being a good person, it'll never happen to me. A 2014 neuroscience study even pinpointed the just-world belief in the brain.[28] Participants with stronger just-world beliefs, as assessed by the usual questionnaire methods, showed greater activation in parts of the brain that are triggered by unethical behaviour* (in this study, a vignette about one colleague blackmailing another with salacious photos from the office Christmas party). That is, although almost all of us, deep down, maintain a just-world belief to some extent, some feel it

---

* Insula and somatosensory cortex, if you're wondering.

so strongly that they have a 'hair trigger' for anything that's just not quite right.

Why do we maintain the just-world belief, despite daily evidence to the contrary? Essentially, according to Lerner, it's what allows us to get up in the morning. We need to believe that if we do our best, then our endeavours – be they professional, social or romantic – will pay off. If this isn't true, if all is chaos, then why bother? Underdog stories are catalysts for human progress, then, because they chime with a belief that is absolutely fundamental to our existence. A world without Leicester Citys and Jyoti Mishras working their way up from nothing to beat the big boys at their own game is simply not a world that we can bear to live in.

8.
SACRIFICE

The sovereign state of Panem, in North America, is made up of twelve semi-independent districts and a central federal district run directly by the government (like Washington DC in the USA). The government, based in the country's main city, known as the Capitol, rules the districts with an iron fist through its army of peacekeepers, and taxes them heavily. Years ago, a thirteenth district rebelled against the Capitol, and managed to get hold of some of its nuclear weapons. A compromise was reached: District 13 would be given its independence and sign a non-aggression pact with the government but, in return, would keep up the pretence of having been defeated and destroyed, in order to discourage rebellion in the other twelve districts. To further discourage rebellion, the government set up an annual competition known as the Hunger Games. Each year, each of the twelve districts would select one girl and one boy – aged twelve to eighteen – to take part in a televised fight to the death. (This echoes the Greek myth in which the city of Athens was obliged to annually send seven boys and seven girls to King Minos, to feed the Minotaur. Theseus, who ultimately killed the Minotaur, volunteered to be one of these tributes.)

We join the story at the start of the seventy-fourth Hunger

Games. Primrose Everdeen, aged just twelve, is selected as the female tribute for District 12. Her older sister Katniss, sixteen, volunteers to take her place, and travels to the games with the District's male tribute, Peeta. They quickly identify their biggest rivals: tributes from the well-off Districts 1 and 2 who have trained for the games since childhood (and are therefore known as 'Careers'). Since the Hunger Games are a big TV event, the contestants are naturally interviewed beforehand. Peeta declares his love for Katniss, but she thinks he's just doing it for the benefit of the cameras: key to success in the games is attracting wealthy sponsors who are allowed to send supplies and equipment, and these would-be sponsors love a romance.

Half of the contestants are killed right at the start, but Katniss narrowly escapes and hides in the forest. All four District 1 and 2 Careers survive too and – worse – Peeta seems to have teamed up with them. While hiding out up a tree, Katniss becomes friends with Rue, the twelve-year-old female tribute from District 11, and the two conspire to drop a wasps' nest on the Careers, killing one of them. The other Careers run off, but Peeta stays to help Katniss after she is stung by one of the wasps. It turns out he was on Katniss's side all along, and teamed up with the Careers only to try to keep them away from her (with all their training, you'd think the Careers would have instantly seen through so transparent a scheme, but apparently not).

When Rue is killed by one of the Careers, her district riots. With the president worried about further unrest, Seneca Crane – who runs the games – decides to change the rules to allow joint winners, provided they're from the same district (viewers, like sponsors, love a romance). Apparently taking the bait, Katniss searches out Peeta – who is dying of blood poisoning – setting up the final showdown between Katniss, Peeta and Cato, the last of the Careers. Cato gets Peeta in a headlock, knowing that

SACRIFICE

Katniss, armed with a bow and arrow, won't risk shooting at him. But clever Peeta uses his own blood to draw an X on Cato's hand. Taking the hint, Katniss shoots Cato's hand, and he lets go of Peeta, who throws him to his death.

Only Katniss and Peeta are still standing, so now what? The joint-winners thing was just a ruse, of course – what sort of message would that send out? – so the lovers are obliged to fight to the death. Peeta asks Katniss to shoot him ('they have to have their victor'), but she has a better idea ('they don't'): both eating poisonous berries and dying together. This turns out to be a double bluff: a competition with *no* winners is even worse than one with two winners. Crane backpedals ('Ladies and Gentlemen, may I present the winners of the seventy-fourth annual Hunger Games!') and, as a reward, is locked in a gilded room indefinitely, with only some of the same poisonous berries for company.

On the tiny screen on the back of the aeroplane seat in front of me, the lovers embrace, and as the cabin crew member clears away the plastic tray, I must admit I have a lump in my throat.

THE MASTERPLOT RECIPE

The recipe for the Sacrifice masterplot calls for just three key ingredients. The first is the most obvious. By definition, a sacrifice entails that our hero must be **prepared to give up something of great importance**. Most often, and most powerfully, this is the hero's own life – in the sense that they literally die saving or helping others. This ultimate sacrifice is frequently found in war films such as *The Last of the Mohicans*, *Devotion*, *Saving Private Ryan* and Sam Mendes's technically incredible *1917*; as well as religious films such as *Jesus of Nazareth* (co-written by

Anthony Burgess), Mel Gibson's *Passion of the Christ* or the long-running hit TV series *The Chosen*.

But, as we will see shortly when we look at real-life cases, even sacrifices that are minor on the global scale – sacrifices of time, independence or happiness – can come at great personal cost, to the extent that one really is, in some sense, giving up one's life. Notice that I worded this key ingredient very carefully: the hero must be *prepared* to give up their own life (or whatever); the sacrifice is by no means diminished if things don't turn out that way. Katniss, as it happens, didn't die in the Hunger Games. But she volunteered to take her sister's place in a fight to the death between twenty-four people, thus signing up for – if not *certain* – then at least *highly probable* death. On this reckoning, Katniss sacrificed her life not once but twice: the second time when she and Peeta resolved to swallow the poisonous berries, hoping – but with no great certainty – that Seneca Crane would stay their hands. Similarly, in the biblical story from Genesis, Abraham makes it clear that he is prepared to sacrifice his son, before God calls it off at the last minute:

> Then God said, 'Take your son, your only son, whom you love – Isaac – and go to the region of Moriah. Sacrifice him there as a burnt offering on a mountain I will show you.' . . . When they reached the place God had told him about, Abraham built an altar there and arranged the wood on it. He bound his son Isaac and laid him on the altar, on top of the wood. Then he reached out his hand and took the knife to slay his son. But the angel of the Lord called out to him from heaven, 'Abraham! Abraham!'. 'Here I am,' he replied. 'Do not lay a hand on the boy,' he said. 'Do not do anything to him. Now I know that you fear God, because you have not withheld from me your son, your only son.'[1]

SACRIFICE

It's a similar story in the final chapter of the Harry Potter saga (*Harry Potter and the Deathly Hallows*). We already know that Harry's mother had sacrificed herself to save him. Now it's Harry's turn to voluntarily take Voldemort's killing curse. He doesn't actually die – due to some rather convoluted technicality involving Voldemort having taken his blood earlier – but he is prepared to, and clearly expects to.

The second key ingredient is where things get a bit trickier: the thorny issue of **obligation**. The secret with this key ingredient – like a pinch of salt – is getting the amount *just* right. If there's too much obligation (like for poor old Isaac), you're not making a sacrifice at all, since you have no say in the matter, though someone else may be (here, Isaac's father). If there's too little obligation (if the Hunger Games were purely voluntary), again there's no sacrifice (anyone who volunteered would be doing so for selfish reasons: whether a death wish or a thirst for glory at all costs). Harry Potter wasn't *forced* to take Voldemort's killing curse, but since he knew that doing so would allow Voldemort to be defeated once and for all, it would have looked pretty selfish if he'd declined. The Sacrifice hero must find themselves in a bind, an obligation sweet spot: they don't *have* to do it, but they *should* do it. Nobody's *making* them do it, but everybody (including the hero themselves) thinks they *ought* to do it.

The third key ingredient is (**pseudo-**)**family ties**. In the prototypical case, our hero is obliged and prepared to give up her own life to save a literal family member (like Katniss and her sister or Lily Potter and her son, Harry). Sometimes, the definition of family member is expanded a bit to include lovers (Katniss and Peeta eventually end up married with children, meaning that her second potential sacrifice is for a would-be future family member). Sometimes, the definition is expanded much more to include those who are not current or future family members in any literal sense, but who are 'brothers' in religion or an army; for example, Private Ryan's 'brothers' Captain John Miller and his men (Ryan's three actual brothers having already been killed in the war). I won't say any more about this last scenario for now, as it will be clearer once we've explored the science behind the story. And, in turn, before we can understand that science, we need to first set aside fictional sacrifice, and look at a real-world example.

STRANGER THAN FICTION

Occasionally, real-world sacrifices are just as dramatic as Katniss's, or even more so: although she is prepared to die in place of her sister, it turns out that she doesn't need to. As we will see later, countless soldiers who put their lives on the line to save their buddies were not so lucky. But as with most things in the real world, stories of sacrifice are more often than not humdrum, even – whisper it – boring. Here's one that is certainly not about to be made into a Hollywood blockbuster.

Irene Hill, known to everyone as Rene (pronounced to rhyme with 'genie'), grew up in the seaside village of Chapel St Leonards, in a house right on the seafront, overlooked by the sand-hills leading to the beach. Rene did well in school and, after graduating, moved to Nottingham to train to be a teacher: just about the best job available to a young woman growing up in 1930s Britain. Living in the shadow of the famous cricket ground, Rene would each morning walk across the Trent Bridge to her college in the city centre. Nottingham was – and remains – a medium-sized British city, but to this young girl from a seaside village it felt like a heaving metropolis. Rene would often see Jesse

SACRIFICE

Boot, the founder of chemist chain Boots, driving around the city in his imported Métallurgique, one of the earliest cars. It all felt like something from the future. And in many ways, it was. It was almost unheard of, in those days, for an unmarried woman to live away from home, and Rene relished her independence.

But not for long. Soon a letter arrived. Rene's father, Will, was seriously ill. Would she move back to Chapel St Leonards to help care for him? She would. Rene took a job elsewhere in the village, working as housekeeper for an accountant known locally as 'Mr A' or 'Mr Tom' (though his name was actually Ernest). Rene's main job was caring for Mr A's sick wife, Margaret (on the side, she volunteered for the Red Cross). In 1940, Margaret died, aged fifty-nine. Mr A asked his housekeeper whether she would like – as she so nonchalantly put it when recounting the story years later – 'to stay on . . . as his wife'. Who could refuse such a romantic proposal? Certainly not Rene. The couple were married in October 1942, shortly before Mr A's sixtieth birthday. Rene, herself just thirty, had married a man not only twice her age, but older than her own father (on whose birthday, incidentally, the wedding fell).

If Rene's parents had any misgivings about the wedding, they kept them under their hats. Her mother, May, kept a diary and noted only that, 'It went off very nicely on Wednesday . . . Rene looked very nice in her deep red satin crêpe dress with navy hat and shoes'. She goes on to talk about the cake ('excellent' despite the wartime shortage of icing sugar) and the refreshments ('Jean [Rene's sister] and I had tea as we don't like coffee'). She mentions her daughter's husband only once, albeit to throw shade on his piano playing: 'Mr Hilson bluntly informed Tom that he could not sing to his playing'. But if she was concerned that her daughter's husband was older than her own, she didn't mention it. The subsequent domestic arrangements were . . . unusual. Rene shared a bedroom of the new bungalow not with her husband, but with her younger sister, Jean. When Mr A felt so inclined, he would leave a note to that effect, finger-written in condensation on the sisters' bedroom mirror, and the new Mrs A would make the short trip to the next room.

The visits were productive. By the time Ernest died in 1958, Rene had raised two children – Tom (named after his father's nickname) and Margaret (after her father's first wife!), aged just thirteen and nine when their father died. Rene had enjoyed little to no help from her husband – as was the case with most fathers then, let alone those in their seventies – but now she had to raise the children alone. She did so uncomplainingly, and the children thrived, passing the eleven-plus and going off to university in Liverpool and Essex to study electrical engineering and politics respectively. After university, Tom Junior settled in the south-east (near Ipswich) and Margaret Junior in the north-east (near Middlesbrough), with Rene – still on the east coast – in the middle. But still her service wasn't complete. Tom and his wife invited Rene to come and live nearby to help look after their children: me and my sister. This she did, until she was too frail to do so any more, and moved into a care home. We'd visit once or twice a week. The funeral, when it came, was a family-only affair. The vicar pronounced her name as 'Reen' throughout.

Although it had its moments of happiness, Rene's life was essentially sacrificed in the service of others. And whether or not she could quite join the dots herself – which became more and more difficult as Alzheimer's set in – on some level she knew it. She never said it flat out, but she danced around it constantly: the feeling that she'd missed out on life, that we didn't appreciate how much she'd given up. And, yes, of course, I have almost no understanding of what life, family or society was like in the middle of the twentieth century, and most of Rene's story comes to me second- or even third-hand. But, on the face of it, it seems to be a quintessential real-life example of the Sacrifice masterplot. Rene **gave up something of great importance** – her own ambitions, her own shot at happiness – for others. The story has just the right amount of **obligation** – nobody *forced* her to care for her ailing father, but it would have somehow been unthinkable to refuse, given the **family ties** at stake (in this case, real, rather than pseudo). So, did she jump or was she pushed? Was her sacrifice freely chosen? And if not, is it a real sacrifice? What exactly is obligation, and how is it created by real or even pseudo-family ties?

I need to speak to an expert.

THE SCIENCE BEHIND THE STORY

The town of Frodsham – like its most famous son, Take That's Gary Barlow – is pleasant, if bland. Like every small English town in 2024, its high street is dominated by discount supermarkets, vape stores and charity shops, where I pick up a pair of drumsticks (Vater 5A wood tips, £2) and a Blu-Ray of *Prometheus* (£1.49). Rock-and-roll and exploring the outer reaches of the galaxy are the last things on my mind as I sip my americano in the town's Costa, formerly – as the banner above the Costa signage still proclaims – a 'Land of Beds'. At ten o'clock on a Friday morning, the cafe is rammed. Retirees, big groups of harassed-looking young mothers, the odd middle-aged man in a suit looking for a socket to plug in his laptop. Literally, the last place on earth you would associate with suicide bombers.

That's what I'm here to talk about. My interviewee – Professor

Jon Cole – ambles in, with his greying blonde hair and beard and gentle south-west accent, wearing shorts despite the summer rain lashing the pavement outside. He orders a double espresso and apologizes for being late. He was, he explains, inviting a colleague to throw him under the bus in discussions with their mutual employer: 'Sacrifice'.

Cole is one of the country's leading experts on suicide bombers, having literally written the book on *Martyrdom*. I wanted to ask him whether 'sacrifice' is the right term, the right narrative framing, for what it is that suicide bombers do. After all, this is hardly the dominant framing in Western media which, understandably, portrays suicide bombers as either evil, fanatic psychopaths or – slightly more charitably – vulnerable simpletons who have been brainwashed. Cole answered my question with a question: what would I do if I found out that somebody had been molesting one of my kids? I'd beat them up, he suggested. (I nodded, while secretly thinking that I'm not really the beating-people-up type, and would be more likely to seek my vengeance through the legal system; but maybe nobody really knows how they'd react until it happens to them.) 'And your wife,' he continued, 'would definitely try to kill them, and you'd have to stop her.' Again, I had my doubts, but I could see what he was getting at. It's a cliché, but all parents would give their lives for their children, whether by accepting a life-long prison term for killing an abuser or – if it were possible legally and medically – by donating a vital organ.

Is giving up your life for your kids a 'sacrifice'? It depends what you mean. On the one hand, giving up your life so that another person might live is the very definition of sacrifice. On the other hand, when we're talking about your own children, your own flesh and blood, 'sacrifice' doesn't feel like quite the right word. Giving up your own life for the good of your

SACRIFICE

children is just day-to-day parenting: doing the school run when you'd rather be having a lie-in; braving a soft-play centre when you'd rather be literally anywhere else in the world; or for the majority of parents worldwide – and putting my first-world 'problems' into perspective – making sure your children are fed while you go hungry. These things don't feel like sacrifices, because we are genetically hardwired to put our children first. We don't feed our children out of altruism, but selfishness: we feel good when they're satisfied, and bad when they're hungry (*actually* hungry, that is, not just whinging for junk food).

We'd all give our lives for our children. But how far does the circle extend? Katniss laid down her life (or, at least, gambled her life at odds of 24:1 against) for her sister. Rene gave up her life (or at least, her career and her independence) for her father, sister, husband, children and grandchildren.

So, what's your price? In response to the question of whether he would lay down his life for his brother, the biologist J. B. S. Haldane (or maybe it was John Maynard Smith, or maybe the story is apocryphal[2]) supposedly said, 'No, but I would gladly give up my life for two brothers or eight half cousins.' Genuine or not, the line contains an important insight. Like your child, your brother or sister shares (roughly) half your genes; your cousin a quarter; your half-cousin an eighth, and so on. Giving up your life for your child is no ask at all; you're genetically programmed to *want* to. But the more distant the genetic relationship, the weaker that genetically based urge.

This explains why bona fide **family ties** are a key ingredient in the Sacrifice masterplot; it's all just biology. Where do pseudo-family ties come in? The answer is that humans *aren't* all biology. We've been able to use culture, law, tradition, society, science and technology to short-circuit these biological urges. For example, (almost?) no man in a modern industrialized country goes around impregnating literally as many women as possible, as he would if he were acting according to 'pure biology' (and, indeed, as the males of many other species do). Society would frown upon him, the law would make him pay child support, science and technology would give him contraception or – failing that – pornography, and so on.

A similar shortcircuiting has happened, Professor Cole told me, with suicide bombers. So tightly are they bound by society, culture, tradition and – yes – religion that they have widened the circle of those for whom they would lay down their life to the 'brothers' who share their cause. It is no accident that religions – including hardly radical ones like the Church of England – encourage their adherents to call each other, and to consider each other, brothers. Nuns are literally called 'Sister', for God's sake.

So, are suicide bombers making a 'sacrifice'? Yes, if we understand a 'sacrifice' not as a selfless altruistic act that you do purely to make *others* feel good, but – like feeding and clothing your own children – one that you do to make *yourself* feel good, by helping your genetic kin. We all broaden our circle a bit – wouldn't you help a distressed stranger in the street, particularly if they seemed at least somewhat like you? – but suicide bombers just broaden that circle more than others.

But – Professor Cole was quick to add – it's not *just* suicide bombers. Medals for gallantry, such as the Victoria Cross and the George Cross in Britain, are given almost exclusively to soldiers who risk their life to save their brothers (yes, soldiers too are encouraged to see each other as 'brothers'). Looking, for example, at the 2001–21 War in Afghanistan, around half of the soldiers who received one of these medals did so posthumously. They really did lay down their lives for their buddies. This is not, I hasten to add, intended to imply any moral equivalence between someone who blows themselves up to kill and injure civilians, and someone who dies trying to save others. The point is that, *from the perspective of the person who gives their life*, the underlying motive of sacrificing their own life to help their broadly drawn circle of brothers is the same.

Thanks to Professor Cole, we've nailed the science behind **pseudo-family ties** as the motivation for being **prepared to give up something of great importance** (often one's own life), but what about the final key ingredient: the sense of **obligation**? Here, we turn to the psychologist Mike Tomasello, who literally wrote the book – well, the journal special issue – on 'The Moral Psychology of Obligation'.[3]

Chimpanzees, Tomasello points out, don't have obligations. They hunt in groups, but if one chimpanzee does a bad job, they don't get chastised or apologize; they don't owe the group anything. And at the end of the hunt, there is no obligation to share the spoils equally – the most dominant male just takes what he wants; maybe there are some scraps left for the others to fight over, maybe there aren't.

At some point in evolutionary time, as the first primates that start to resemble modern humans emerged, those primates realized that hunting is more efficient if the members actively cooperate, assigning one another specific roles: you stalk the prey, I scare it out of the bushes, he throws the spear at it and so on. The quid pro quo is that we all get an equal share of the meat afterwards. With these first glimmers of cooperation, we also see the first glimmers of obligation. If you don't hold up your end, if I scare the prey out of the bushes, but you fail to throw the spear at it, you've let me down; I'm entitled to chastise you, and maybe even withhold some or all of your share of the kill.

This is the first part of the story, but we're still talking about obligations that arise from more-or-less explicit commitments made to just one or a handful of other people. How do we get from here to unspoken obligations that extend to people we have never met? Tomasello's answer is that around 150,000 years ago, early humans started to live in much bigger groups. This meant that it was no longer possible, as it had been before, to recognize all of your group members by sight. This ability is of vital importance, as members of other tribal groups are likely to be armed and dangerous.

The solution is shared culture: shared songs, dances, clothes, ways of talking and so on. Individuals are now *obliged* to dress, act and behave in certain ways. Anyone who refuses is putting the group at risk by chipping away at the system that protects everyone from dangerous outsiders, and therefore must be ostracized or otherwise punished no matter how – on the surface – minor their transgression (e.g. failing to master the tribal dance). There are practical benefits, too, to having everyone sing from the same hymn sheet. If a group of strangers are hunting, it will be much easier for everyone if they all turn up with a shared understanding of 'the way we do it'. And, just as with the much-smaller hunting groups of earlier in evolutionary time, everyone who has played by the rules can expect their fair share of the spoils at the end of the hunt. It is now in everyone's interest not just to stick to the rules themselves, but to chastise those who go off piste.

Carry this process forward for 150,000 years, as groups get larger and larger, and the shared cultural norms – 'the way we do it here' – get more and more sophisticated, and we reach modern-day societies. As long as they play by the rules, we are obliged to treat other members of our society fairly and equitably, and to play by the rules in our dealings with them, even if they are complete strangers. Even young children know this.[4]

SACRIFICE

If two four-year-olds who have never met play a game (as part of an experiment) in which they must work together to win sweets, they almost always share the sweets equally afterwards. If the experimenter fixes the game so that one child wins more sweets than the other, *both* children protest, and the one who has got too many will normally hand over as many as it takes to make it fair. And it's not just for important things like winning sweets. If children are taught a new game with completely arbitrary rules, they will protest like mad if an experimenter plays the game in the 'wrong' way, even if it doesn't matter in the slightest. Even four-year-olds know that we're all obliged to follow norms and conventions, no matter how arbitrary.

That, then, is the second key ingredient of sacrifice: **obligation**. But how do we get from here to the scenario where someone feels obliged to give up their life? We have some obligations to strangers – like sharing sweets fairly – but these obligations don't include dying for them. Professor Jon Cole, in our discussion of suicide bombers above, has already given us most of the answer: we are genetically programmed to be prepared to die for our children and other close kin; and some groups piggyback off this preparedness by persuading their members that they are – in effect – each other's family. The final piece of the jigsaw, then, is understanding just how these groups go about building such effective pseudo-family ties.

The leading theory, as set out by Oxford anthropologist Harvey Whitehouse,[5] is that 'fusion' – merging one's personal identity with the group – happens when people undergo a transformative experience together, particularly when that experience is traumatic. Whitehouse discusses studies conducted with survivors of the high-profile terrorist attacks in London, Madrid and New York, showing that recalling and discussing their experiences increases their willingness – in surveys afterwards – to die for their country. Researchers have even started to uncover the biological mechanisms behind this fusion. During the 2014 World Cup in Brazil (won by Germany), researchers hooked Brazilian fans up to equipment that measured their heart rate and levels of cortisol – a hormone related to stress – as they watched live matches. They found that the higher their heart rate and cortisol levels during the matches, the more 'fusion' with the group the fans reported.

This type of study, of course, measures only low-level trauma (losing a football match) and hypothetical levels of fusion (as measured by a questionnaire). Whitehouse himself therefore set out to interview Libyan civilians who took up arms, many for the first time, against their government in the Arab Spring of 2011. This was a highly traumatic experience – all of those interviewed by Whitehouse had lost friends and family and had been at severe risk of death themselves. Whitehouse asked each of them, if you had to choose just one group with whom you are most highly 'fused', who would you pick? Most of the insurgents picked fellow fighters over their family. Importantly, the same was not true for volunteers who didn't fight themselves, but just provided logistical support to those who did: these 'control group' participants tended to pick their families. It seems that it was going through the traumatic experience of almost dying in battle that 'fused' the group members to one another. Whitehouse found the same pattern in interviews with American Vietnam veterans: the worse the ordeals they had endured, the more they were willing to make financial sacrifices for veterans in need. The same was also true for college fraternity and sorority members who had undergone unpleasant hazing rituals, and for Brazilian jiu-jitsu practitioners who had run a painful belt-whipping gauntlet.

SACRIFICE

This research begins to answer a question that has long troubled anthropologists: just why do so many tribes around the world (as well as street gangs and even university fraternities) have painful 'rites of terror' that involve beating, whipping, mutilation, removal of the fingernails, sleep deprivation, starvation and – if you can imagine it – worse? Ironically – given that the tribe is ultimately the one meting out the punishment – the answer seems to be that going through these traumas together binds the members of the group so that they really do come to consider each other as family (or, as we saw with the Libyan insurgents, as closer than family).

It has been a long journey, but we are finally getting to the core of the science behind the Sacrifice masterplot. Sacrifice involves giving up something important – often one's own life – out of a sense of obligation to others, arising from a notion that those others are family; either because they *really are* family, or because the individuals concerned have formed family-like ties through religion, war or shared trauma.

Although few of us have ever stopped to think about the defining characteristics of sacrifice, deep down, we all know this. How could we not, when just about all of the world's major religions, which until recently dominated society, culture and the law worldwide (and, in many places, still do) place sacrifice front and centre? The central tenet of Christianity is that Jesus sacrificed his life for us. We have already seen the importance that Christians place on the story of Abraham's (would-be) sacrifice of his son. What you may not know – if, like me, you grew up in a broadly Christian culture – is that this event is even more important under Islam, which commemorates it with Eid al-Adha, 'the feast of the sacrifice', the largest of the major holidays. Martyrdom is fundamental not only to Christianity and Islam but also to Judaism, Sikhism, Jainism and the Shinto and Baha'i faiths.* Whether or not we stop to reflect on it, then, all of us have a deep intuitive understanding of the Sacrifice masterplot. Sometimes, as we will see shortly, we fail to put this knowledge to good use, with potentially dire consequences. And when I say dire, I'm talking about the imminent extinction of our species. But sometimes, we apply this knowledge in the service of truly heroic endeavours . . .

UNDER THE INFLUENCE

In early 1945, as the Allies closed in on Hitler, a gruesome experiment was reaching its peak. Thirty-six men had been starved to the point of losing a quarter of their body weight, and were suffering severe physiological and psychological breakdown. But this experiment was not being conducted by Josef Mengele and the Nazis, and the men were not prisoners of war. Incredibly, these men had volunteered to be – as they called themselves – 'guinea pigs' in what came to be known as the Minnesota Starvation Experiment.[6] The purpose of the experiment (a literal Hunger Game) was to investigate how best to treat rescued prisoners who had undergone starvation, as in the prison camps of Europe; a subject about which, pre-1945, almost nothing was known. But before the scientists could investigate how best to *treat* starvation, they needed to induce starvation, under controlled experimental conditions.

You might have thought that the researchers would struggle to find enough willing participants, but no. Estimates vary, but

---

* Amongst the world's major religions, the only ones that do not emphasize sacrifice – in the sense of laying down one's life – are Hinduism and Buddhism; though of course these religions still emphasize the importance of serving others.

historians agree that at least 200 men, possibly as many as 400, put themselves forward. And these men were not dupes. Compulsory intelligence tests taken at the start of the study revealed that these volunteers had IQs in around the top 5 per cent. Neither were they old men who had given up on life – the average participant was twenty-five years old and in peak physical condition, weighing just 73 kg – pretty lean for men with an average height of 5'10".

Who were these volunteers? When the USA entered the Second World War, all men aged between twenty-four and forty-five were required to register for the draft. Around a fifth of 1 per cent of those eligible were 'conscientious objectors' (COs) who refused to fight. Although many were mocked as 'cowards', COs weren't just people who didn't *want* to fight – to qualify, it was necessary to demonstrate a genuine moral or religious conviction that prohibited it. In practice, this meant that the majority of COs were members of 'peace churches': Friends (Quakers), Mennonites or Brethren. The government didn't want the public to think that these COs were being given an easy ride, and so, with the agreement of the Churches, set up the Civilian Public Service, under which COs would be required to do 'work of national importance'.

Often, this work was unpleasant or dangerous, or both. Some COs fought forest fires; others worked in psychiatric hospitals. Some – like the thirty-six who took part in the Minnesota Starvation Experiment – volunteered for medical research. The key word here is *volunteered*. Unlike the truly horrendous Warsaw Ghetto study, in which the Nazis restricted prisoners to less than a tenth of the standard daily calorie allowance, the participants of the Minnesota Starvation Experiment were volunteers. And, remember, they were high-IQ men, all of whom had to pass psychological screening tests. They knew what they were letting themselves in for. Why did they do it? Out of religious conviction, and a sincere desire to help their fellow man. The recruitment advert to which the volunteers responded was perfectly pitched: 'Will you STARVE that THEY be better fed?' it asked, below a picture of starving children, deliberately echoing a Bible verse that lies at the heart of the Mennonite and other peace churches: 'For I was an hungred, and ye gave me meat: I was thirsty, and ye gave me drink: I was a stranger, and ye took me in' (Matthew 25:35; King James Version).

What is it like to starve? I can't even begin to imagine. It just so happens that, when I first heard about the Minnesota Starvation Experiment, I had recently started a diet. At 5'11", my 84 kg made me officially overweight (according to the NHS BMI calculator), so I set myself the target of getting down to 80 kg, which would place me in the 'healthy weight' category, albeit right on the borderline. Today, as I look over this chapter for the last time before sending it to my editor, I reached my target. This has taken five weeks of replacing breakfast and lunch with meal replacement drinks and – other than the occasional apple or orange – a complete ban on snacks and non-diet drinks. As you'll appreciate if you've ever tried a similar diet, I'm hungry all the time, usually grumpy, and my breath stinks (something to do with 'ketosis', apparently). But I haven't even got close to scratching the surface of what the Minnesota volunteers went through. So far, I've lost around 5 per cent of my body weight. The Minnesota volunteers had to lose *25 per cent*.

You may be surprised to learn that most of the men achieved this target on a relatively generous-sounding allowance of 1,500 calories per day, which is more or less my current regime (in the UK, the standard NHS recommendation is 2,500 calories a day for men and 2,000 for women). The difference is that the Minnesota volunteers were also required to walk twenty-two miles each week (that's doing a 5k every day) and – here's the kicker – to keep this up for twenty-four weeks. Cigarettes, water and gum were unlimited, as was – at first – black coffee, though this later had to be restricted to nine cups a day, as many of the men began to drink it in truly alarming quantities.

The impact of the weight loss was much worse than even the experts had anticipated. Of course, by 1945, there had been many terrible cases of starvation, but it was generally thought that the worst effects were from the uncertainty of war and famine rather than the starvation itself. In contrast, these volunteers 'were under constant medical supervision, and knew that they would be taken out of the experiment if anything went wrong', wrote the authors of *Men and Hunger*,[7] a short summary of the findings rushed out quickly after the study had finished, as a guide for aid workers. 'They were not bombed. They knew beyond a shadow of a doubt that more food would be given to them at the end of the six months of starvation – their food rations were not dependent upon the vicissitudes of politics'.

It didn't matter. The physical changes were bad enough. The typical pulse rate halved (the lowest recorded was 28 bpm, as compared to a typical 60–100 bpm). White blood cell counts decreased by 30 per cent. Hair and nails fell out, cuts and bruises healed slowly. The volunteers were cold all the time, wearing multiple layers, lying in the sun and asking for their food and drink to be served extremely hot. Fatigue set in – one of the volunteers, Sam Legg, reported:

> I was walking along with my buddy, it was deep into the semi starvation and we were tired. We would look for driveways when we got to a cross street so we wouldn't have to walk up one step to get from the road to the sidewalk.

But perhaps even worse than the physical changes were the psychological changes. The men became obsessed with food. One man collected a hundred cookbooks: 'Stayed up until 5:00AM last night studying cookbooks. So absorbing I can't stay away from them'.

Another, long after the study had ended, carried a photograph of a cinnamon roll in his wallet at all times. The men lost all interest in sex. Some diluted their food with water to make it go further and take longer to eat (which they called 'souping'). Others seasoned their food until it was encrusted with a thick layer of salt and pepper. They became 'senselessly irritable, particularly when . . . watchin[g] some of the bizarre eating habits of others':

> I find I am becoming more and more frank about showing my emotions . . . particularly at the table. I got up and left the table, telling [a fellow volunteer] that I did not appreciate his licking his plate so noisily. I told him that he sounded like a damn cow.

Some began to hallucinate. Sam Legg, quoted above, was chopping wood, and suddenly imagined slicing through meat. He wasn't able to say whether he'd done it on purpose or not, but he chopped off three fingers. Legg begged – successfully – to be allowed to continue with the study. One man, Franklin Watkins, was kicked out after dreaming about cannibalism and – after his money was taken away – shoplifting food.

Despite the horrendous suffering the volunteers went through, the experiment was a success. The researchers uncovered a wealth of valuable information concerning not just the effects of starvation but – more usefully – how to help people recover afterwards. First, worried about the potential dangers of too-much-too-soon in the recovery phrase, the study director, Ancel Keys, started the men on 2,200 calories a day: about what you or I would eat on a typical day. Surprisingly, on this regime, the volunteers didn't put on any weight at all. It turned out to be necessary to give them around 4,000 calories a day for several months (though, when given the opportunity, most consumed around 10,000 calories a day for a few weeks before levelling off). Another important insight was that, counter to what had been hoped, there was no magic bullet: it doesn't help to put starving people on a special high-protein diet, to give them vitamin or mineral supplements or to feed them through a nasogastric tube; it's just the calories that count. Another important – though problematic – finding was that, in the recovery phase, the body builds fat quicker than muscle, to the extent that – in order to repair their damaged muscles and organs – the volunteers typically had to eat to the point that their abdominal (belly) fat was 40 per cent higher than it was before the study.

Why was the experiment such a success? Because the heroic volunteers – who could, remember, have quit at any time – were living out a narrative that ticks all the boxes of the Sacrifice masterplot. They were, it goes without saying, **giving up something of vital importance** – their physical and psychological wellbeing; but they were doing so out of a sense of moral **obligation**.

> 'It's colored my whole life experience and it was one of the most important things I ever did.' – Wesley Miller

> 'I think probably most of us are feeling we did something good and we're glad we did it and that helps us live a better life.' – Sam Legg

When Legg, after chopping off his fingers, was threatened with being dropped from the study, the response he gave is dripping with obligation:

> Keep me in it, for the hungry. For the rest of my life, people are going to ask me what I did during the war. This experiment is my chance to give an honorable answer to that question.

Where does this obligation come from? From **pseudo-family ties**. From the fact that, both during the experiment and in their persecution as 'draft-dodging cowards', the terrible experiences endured by these men brought them together as family; from the religious conviction that allowed them to extend the circle of 'kin' to all mankind, to – in the names of their Churches – their 'Friends' or 'Brethren'; in short, from the Sacrifice narrative.

PLOT TWISTED

When put to the right use, then, the Sacrifice masterplot can save lives. When it is put to the wrong use, on the other hand, it can cost lives: feasibly, the life of every man, woman and child on the planet. As I write this, we've just seen the hottest month globally since records began. We're on course to break the record again this month, and by the time you read this, it will no doubt have been broken again. I don't know how to put it any more clearly than this: WE ARE ALL GOING TO DIE. Everyone knows this. Yet, as we will see in more detail in the next chapter, nobody is really doing anything about it.

Why not? One of the biggest problems, as I see it, is that climate change has been framed as a Sacrifice narrative. Let's see what sacrifices you make.[8] First, I'd like you to give yourself

SACRIFICE

a score for how environmentally friendly you are in your day-to-day life. We'll use a 1–5 scale where 1 = *I don't make any effort at all to help the environment* and 5 = *I do everything I possibly can*. This is your 'intentions' score. Now let's work out your 'actions' score. How often do you do each of the actions listed below? Score each from 1–5 (1 = never, 5 = all the time), then take the average (i.e. add up your scores and divide by 10).

a) switch off lights
b) put on a jumper rather than turning up the heating
c) avoid buying something due to packaging
d) take your own shopping bag
e) use public transport
f) walk
g) buy recycled toilet paper
h) avoid taking flights
i) avoid leaving your TV on standby
j) turn the tap off when brushing your teeth.

If you're like most people, your *intentions* score (for most people around 4), is a point or two higher than your *actions* score (for most people between 2 and 3). We like to think we're doing our bit, but, actually, most of us are not; or at least a lot less than we think we are.

Before you start to feel guilty, you should know this: the personal-sacrifice narrative for addressing climate change didn't develop organically from the ground up. No, this perversion of the Sacrifice masterplot was done quite deliberately. I'm sure you've heard of the idea of your 'carbon footprint' – the amount of carbon that you produce while going about your everyday life; everything from burning petrol in your car to drinking wine that has been flown in from Australia. What you probably don't know is that the concept of the carbon footprint was invented by an advertising firm working for – you guessed it – Big Oil, specifically BP (formerly British Petroleum).[9] This concept is very useful to the fossil fuel companies, as it focuses the attention squarely on individuals ('What are YOU doing about the comet problem?'), letting the companies off the hook (that personal-sacrifice narrative again). The truth is, the only reason that you, as an individual, have a carbon footprint *at all* is that everything from the car that you drive to the plane that imports your wine is powered by burning gas and oil sold by BP and their ilk. If BP and the others were to switch to 100 per cent renewables, your carbon footprint would rapidly shrink to zero.

So, will they? Hollywood doesn't think so. In fact, in the one impending-comet movie in which humanity *isn't* saved, *Don't Look Up*, the job of destroying the comet is outsourced to a private company, who plan to mine it for rare-earth elements (BASH Cellular, the billionaire CEO of which is a leading donor to the political party currently in government). Needless to say, this does not go well.

While it might be a convenient framing for BP and their ilk, everything we have learned about the Sacrifice masterplot tells us that asking everybody to make small daily sacrifices is a *terrible* way to fight climate change. The Sacrifice masterplot – whether it's the Minnesota Starvation Experiment, *The Hunger Games*, *Harry Potter and the Deathly Hallows* or Rene Ambridge – involves giving up something big and important, often one's own life, out of a sense of obligation to others, amplified by the feeling that those others are 'family'; a feeling that has arisen from some shared traumatic experience. The 'sacrifice' approach to saving the planet has literally none of those key ingredients.

First, the sacrifices that people are asked to make are tiny and trivial: switching off lights, taking your own bag with you when you go shopping and so on. Even the biggest ones like going vegetarian or giving up flying don't begin to compare with sacrifices like starving yourself to the point of hallucination, let alone giving up your life for a cause. The problem is not just that personal sacrifices are ineffective, though they probably are – one recent report found that over 70 per cent of global emissions since 1988 can be traced back to just a hundred companies.[10] The problem is that tiny sacrifices like turning the tap off when you brush your teeth are too trivial to make people buy into the sacrifice narrative in a meaningful way. People don't neglect to turn the taps off because it's too much effort, but because it's too *little* effort. For the Sacrifice masterplot to work its magic, we need to feel that we're doing something big and important.

The second reason that the Sacrifice masterplot risks dooming us all when applied to climate change is that it doesn't tick the box of **obligation**. The social mores of the 1930s *demanded* that my grandma set aside her own career ambitions to care for her sick father. Katniss's love for her younger sister *gave her no other choice* but to take her place in the Hunger Games. The suicide bomber's/kamikaze pilot's/ heroic soldier's/Libyan rebel's/ Minnesota guinea pig's commitment to the cause meant that he or she felt *obliged* to make their sacrifice. This sense of obligation is entirely missing from the sacrifices that we are asked to make to save the planet. In fact, the worst part of it is that we are not even really 'asked' to make these sacrifices at all. So little obligation do we feel, and so obsessed are we with personal autonomy, that these sacrifices are offered as mere *suggestions* that some of us might like to follow *if we feel like it*. But if we don't, no problem, it's all about personal responsibility and personal choice. The sense of obligation is missing entirely.

Why is this sense of obligation missing? This brings us to the third and final way in which this approach to saving the planet fails to tick the boxes of the Sacrifice masterplot: we feel no sense that the necessary sacrifices are made for the good of our genetic kin, or for non-kin with whom we have fused or bonded through some shared traumatic experience. The irony is that, in the long term, the sacrifices that we need to make *are* for our genetic kin: if not our children, then perhaps our children's children, and certainly our children's children's children. But it's all too far in the future, too abstract, too diffuse, too theoretical. Would you give up your car for ever to save your children's lives right now? Of course – who wouldn't? Would you give up your car for ever to – potentially, as long as everyone else does it too – save the life of your children's children and a bunch of other people's children? What's that? You'll get back to me? The irony is that as climate change takes hold, and large groups of people go through literally hellish experiences together, they will – at least if Harvey Whitehouse's theory is correct – become more willing to make sacrifices for one another. But by then it will be far too late.

HAPPY ENDINGS

The Sacrifice masterplot is one in which our hero is prepared to – and often does – give up something of great importance, often their own life, out of a sense of obligation to others. This obligation arises because the people for whom the hero is making the sacrifice are either literally genetic kin – in which case the obligation has a purely biological origin – or pseudo-family with whom the hero has bonded over war, religion or some shared traumatic experience (an example of society co-opting or short-circuiting biology).

SACRIFICE

Humanity's relationship with the Sacrifice masterplot is – probably more so than for any of the other masterplots in this book – a fraught and complex one. Is Sacrifice a catalyst for human progress, or a brake on it?

Often, it's a matter of perspective. Let's go back to Professor Jon Cole's example of suicide bombers. As Andrew Silke asks in 'Understanding suicide terrorism: Insights from psychology, lessons from history',[11] given the opportunity, would you put on a suicide vest and kill Hitler in 1943? The question wasn't a hypothetical one for Rudolph-Christoph von Gersdoff, who indeed chose to put on the vest and make an attempt on Hitler's life. With the fuse lit, von Gersdoff got close enough to Hitler, but unfortunately the bomb had a ten-minute fuse, and Hitler – quite unexpectedly – left the event (a museum exhibition of captured Russian weapons) after only a couple of minutes.

Had von Gersdoff succeeded, the Nazis would have portrayed him – Silke points out – as 'a brainwashed fanatic, coerced or "radicalised" into killing himself'. Yet you will almost certainly agree with me that his actions were 'reasonable and acceptable, and his willingness to sacrifice himself in the attempt an indication of enormous personal courage'. We could point to many similar examples from history. Was the suffragette Emily Davison, who threw herself in front of the King's horse at the Epsom Derby race in 1913, an unbalanced suicidal fanatic (as she was largely portrayed at the time) or a brave freedom fighter (as most people would regard her today)?[12] Ultimately, whether sacrificing oneself for a political cause is a catalyst for or a brake on human progress depends on your personal view of the particular political cause, and whether or not that cause represents human progress.

Lower-level sacrifices, such as that made by my grandma, are similarly fraught. If we, as a society, encourage this type of sacrifice, is that progress, or the opposite? Again, it's a matter of perspective. Some would say that her actions were noble, and gave her life meaning. Others would say that she did a great disservice to herself, and perhaps even her family, by not pursuing her own happiness, and that her life should not be held up as a model for others. And, as we have just seen with the example of climate change, an inappropriate Sacrifice narrative can be a catalyst not for human progress, but for human extinction.

Yet the example of the brave volunteers in the Minnesota Starvation Experiment tells us that, given the right circumstances, the Sacrifice masterplot can indeed serve as a catalyst for human progress. Because, make no mistake, this experiment really did drive human progress in terms of our scientific understanding of the effects of – and treatment for – starvation; insights that are still used today.

These days, most societies, particularly in the West, place great emphasis on autonomy, personal happiness, living your best life. Sacrifices like those made by my grandma are most definitely out of fashion; and, on balance, that's probably a good thing. But we risk throwing out the baby with the bath-water. Most experts think that the Minnesota Starvation Experiment would not be possible today, since no university ethics committee would permit such a study. If they're right, then – in my view – it's a great pity. Even if they were mocked at the time, history reserves its greatest admiration for those who sacrificed something important for a noble cause. Millions of consenting adults have found meaning in their lives and deaths and even – like the suffragettes and Minnesota volunteers – advanced human progress, by living out the Sacrifice masterplot. Who are we to stop them?

9.
HOLE

Los Angeles, 2029. Machines scatter searchlights, and laser beams fire across a post-apocalyptic landscape. Their goal is to eliminate humankind.

Los Angeles, 1984. Sarah Connor, a harassed waitress, knocks over a glass of water. As she clears it up, a kid scoops some ice cream into the pouch of her apron. 'Look at it this way,' says a fellow waitress, 'in a hundred years, who's gonna care?'

A bodybuilder – who turned up on the street naked as if dumped there from outer space – has been looking up Sarah Connors in the phonebook and killing them (thankfully, after stealing some clothes off a bunch of ne'er do wells). Meanwhile, another young man – attractive, but boy-band type; not a bodybuilder – who also turned up naked on the street, has been dreaming about fighting against machines in some kind of futuristic war.

Our Sarah Connor sees a TV report about the murders of the other Sarah Connors, the police having briefed the media after trying and failing to contact her. She flips through the phone book and finds that she is the third and last Sarah Connor listed. She has well and truly fallen into a hole.

What should she do: run away? Call her flatmate? Not answering. Call the police? OK, they're on the way. But what to do in the meantime? Hide? Have a seat? Sure, it's a busy bar. But over there is that weirdo who's been following me (the boy-band one). And I don't much like the look of this bodybuilder one who's – wait – *pointing a gun at me!* Now the boy-band one has shot the bodybuilder one through the window. 'Come with me if you want to live,' he says.

She does. The two try to escape in a car, but the bodybuilder is still after them, punching through their windscreen, despite being on fire. Clearly, we're dealing here with something that is not of this world. Kyle Reece (sweet like the candy) explains to Sarah that their pursuer is indeed not human. He's a Terminator.

Phew! We need a breather, in the form of the 'B story'. You know, in fiction, the subplot that seems tangential to the main, 'A' story but, as we will soon see, is often anything but. We hear from Kyle about Sarah's unborn son John, who leads the resistance against the machines in the future. That's right – the machines sent the Terminator back in time to kill Sarah before she can give birth to John. As for the love interest part, usually a central part of the B story, we're not yet told who John's dad will be, but perhaps we're starting to get an inkling.

We're now at the exact midpoint of the story and our heroes have a temporary escape: Kyle and Sarah are arrested and locked away, safe from the Terminator. But the respite is only temporary: when the Terminator comes looking for Sarah at the police station, the desk sergeant tells him she's making a statement and it may take a while. 'I'll be back,' he says. Of course, he soon is, armed to the teeth.

Sarah and Kyle escape to a forest first, then to a cheap motel, where they while away the hours making pipe bombs: moth balls, corn syrup, ammonia. It's not all bad, though, as they do find the time, and the mood, to conceive John (you know, her son from the future, from the B story).

The Terminator now tracks down Kyle and Sarah (which he does cleverly by impersonating her mother on the phone), setting up the finale. Our heroes launch their pipe bombs at the Terminator, then his vehicle. But there's a setback: even though the explosion was enough to kill Kyle (sorry, mate) and split the Terminator in two, his top half is still pursuing Sarah. She takes a clue from the vision of the future Kyle set out for her (and that we saw in the B story). What can kill a machine? That's right – another machine. She lures the Terminator into a hydraulic press and crushes him.

Our final image is of Sarah hiding out in Mexico, pregnant with John. Maybe humankind won't be eliminated after all . . .

THE MASTERPLOT RECIPE

The first and most obvious key ingredient of this masterplot recipe is the protagonist falling **into the hole**.[1] This ingredient must be added towards the start of the story, usually at the climax of Act I. Only very occasionally, of course, does the protagonist fall into a literal hole (as in the 2001 and 2009 movies *The Hole*), but the metaphor captures the fact that the problem must be important, instantaneous and involuntary. Often, the problem is on an epic scale (*Terminator*, *The Day After Tomorrow*, *Armageddon*, *Deep Impact*, *Don't Look Up*, *Snakes on a Plane*) but it can relate to an individual (*Die Hard*, *Speed*, *The Wizard of Oz*), a domestic problem (*Sleeping with the Enemy*, *Gone Girl*), a spy problem (*North By Northwest*, *The Bourne Identity*, *Paycheck*) or a personal crisis of identity (*Inside Out*, *The Lego Movie*, *Turning Red*). If the problem isn't instantaneous (or, at the very least, an instantaneous realization of a looming problem), and instead the hero is just generally in a bad way, we're instead into the realms of the Underdog masterplot. If the problem isn't involuntary – if the hero *chooses* to take up arms when they could have just stayed at home – then we're into the realms of the Monster masterplot. No, in a Hole story, our hero is minding their own business when suddenly – wham! There's a bomb on the bus! A meteor headed right for Earth! Snakes on the plane! A Terminator from the future out to kill you! Suddenly, we're not in Kansas any more.

Fast-forwarding to the end of the story, the climax of Act III, we have the second key ingredient – our hero clambering **out of the hole**. Like the first, this key ingredient is non-negotiable. If our hero dies in the hole, we're following a different recipe altogether (almost certainly Icarus). Like a five-spice mix, this ingredient is itself made up of five sub-ingredients. There's quite a bit of flexibility here. Unlike **out of the hole** itself – which is obligatory – each of these five sub-ingredients is optional; but once I've pointed out the mix, you'll notice its distinctive flavour cropping up at the end of a Hole story more often than not.

Out of the hole starts with *making preparations*. If there's more than one protagonist, this literally involves getting everyone together for the big escape (in *Terminator*, that's Kyle and Sarah). If there's just one protagonist, now is the time for them to gather their thoughts, to get all of their ducks in a row. Next, we see our heroes *executing the plan*; or at least trying to. If this is a classic five-point ending, this plan immediately goes awry: *the high tower surprise*. Kyle and Sarah's plan to kill the Terminator with a pipe bomb literally backfires when the bomb kills not the Terminator, but Kyle. It's time for the (remaining) heroes to *dig deep down* and come up with a new plan: crushing the Terminator in a hydraulic press. Invariably, this plan is the right one (*three* would be pushing it!) and the heroes *execute the new plan* (and, here, the Terminator) to perfection. Phew!

HOLE

In between falling **into the hole** and clambering **out of the hole**, almost anything can happen, with the remaining three ingredients – at least in principle – purely optional. In practice, however, it is all but unheard of for a Hole story not to include at least one of these optional ingredients; and the majority will include all three.

As we just saw, when the initial plan to escape the hole is thwarted, our protagonist(s) must *dig deep down* and find an alternative solution to the problem. But where does that solution come from? More often than not, from the B story. Two hours of watching someone trying – and, for the most part, failing – to climb out of a hole is relentlessly grim. The solution that most screenwriters hit upon is to introduce a B story (normally towards the start of Act I): a mini plot arc that is completely irrelevant to the main story. Or so it seems. We'll meet a new character or two, one of whom might well be a love interest for the main character; we'll flashback (or forwards, or sideways) to a completely different setting. At first, this B story seems completely irrelevant; but suddenly – just when our hero is required to dig deep down – the A and B plot lines intertwine to yield a solution. Terminator combines the *love interest* and *new characters* tropes – sort of – as Kyle talks about Sarah's unborn son John from the future, leaving us to infer that Kyle is Sarah's love interest (and vice versa). It's an insight from this story – that a machine is the way to kill a machine – that ultimately provides the solution when Sarah is required to dig deep down at the climax of the story.

Similarly, in *Home Alone* – which, tonally, is almost the exact opposite to *Terminator*, but follows the same Hole recipe – Kevin is ultimately saved from the bungling burglars against whom he has been defending his home by 'Old Man Marley', a neighbour who – as we learned in the seemingly irrelevant B story – reputedly killed his family with a snow shovel; the same shovel that he now wields against the burglars.

The second and third optional ingredients of the Hole masterplot are mirror images of one another. The first, which usually comes right in the middle of the story, is the **false dawn**, where our heroes gain a temporary reprieve. Kyle and Sarah are in jail, where they are safe from the Terminator. In *Home Alone*, Kevin has scared off the burglars by using the soundtrack of an old movie and some firecrackers in a saucepan to make it sound like his house is occupied by gangsters. In *Deep Impact*, astronauts land on the comet that's headed for Earth, plant nuclear bombs at its centre and successfully detonate them.

Of course, the reprieve is only temporary: the Terminator breaks into the jail; the burglars break into the house; the comet breaks in two (with both pieces still heading for Earth). This precipitates the **dark night of the soul**: a scene in which the main character despairs that all seems lost; that they will never escape the hole. Kyle and Sarah hide out in a dingy motel. Kevin McCallister takes the soul-searching all too literally and goes to a church service. A last-ditch effort to divert the comet fails, and our attention turns to clamouring over who merits a place in the government's underground shelters. Although it's technically optional, most Hole stories will feature a dark night of the soul moment precisely because, as the (technically inaccurate) saying goes, 'the darkest hour is just before the dawn'. It's only by reaching their lowest ebb that our hero is able to summon the inner strength ('This is my house. I have to defend it') to finally struggle out of the hole.

STRANGER THAN FICTION

Do you remember when five people were killed in Saraburi, Thailand, when a bus crashed into a flyover next to a shopping mall? Almost certainly not: in most countries, the tragedy didn't even make the news. But do you remember when a Thai youth football team, along with a coach, were trapped in a water-filled cave for almost two weeks? Almost certainly you do, even though the death toll – two rescue divers – was less than half of that of the bus accident. It turns out that the Hole masterplot recipe is so popular and powerful that even real-life events are most likely to capture the public imagination when they slot neatly into a *Hole*-shaped, uhhh, hole.*[2]

The Tham Luang cave rescue captured headlines worldwide because it could have come straight off the page of a Hollywood screenwriter looking to his next Hole-recipe blockbuster. The unfolding drama was played out on the evening news around the world. More than 100 divers, 2,000 soldiers and 10 police helicopters were on the scene, as more than a billion litres of water were pumped out of the cave. The story had all of the key ingredients; even a B story (or, excuse the pun, sub-plot)

---

* As I was finishing the first draft of this chapter, the news was dominated by the on-the-face-of-it similar case of the *Titan* submersible attempting to reach *Titanic*, with one news channel even featuring a countdown clock which showed the time until the air ran out. This turned out to be even more ill-advised than it at first appeared, as it became apparent that the submersible had already imploded before the countdown began. As some Twitter commentators ruefully noted, this incident attracted far more coverage than the regular death at sea of migrants in Europe, including 600 in Greece just the week before. They have a point but, again, the sad truth is that the *Titan* tragedy attracted far more attention precisely because it fits (or at least initially seemed to fit) this Hole template.

centred around Elon Musk, who volunteered a miniature submarine, then took to Twitter to slam as a 'pedo guy' a British cave diver who claimed Musk's plan had 'absolutely no chance of working' (the diver later lost his $190 million defamation lawsuit, with Musk arguing that the insult was common in his native South Africa).[3] The finale featured a real *high tower surprise* moment, when at first it proved impossible to squeeze the boys through the fifteen-inch gap created by the rescuers, who had to come up with a new plan: taking the air tanks off the boys' backs while they scraped through.

So determined are we to attach one of life's predictable masterplots to real-world events, that only those events that conform readily to one of our eight masterplot recipes make a dent on the public consciousness. In fact, some real-world events follow these masterplot recipes so exactly, it's almost impossible for them not to become Hollywood blockbusters . . .

In 1985, two young Brits, Joe Simpson and Simon Yates, set off for Peru, aiming to be the first to climb the west face of the 21,000-foot Siula Grande mountain. They were accompanied by Richard Hawking, a non-mountaineer friend who had tagged along to mind the base camp.

Joe and Simon successfully climb to the peak and start to descend. Then comes the almost literal fall **into the hole**, when Joe falls and breaks his leg. Usually, a broken leg in such a remote area is a death sentence, but disaster seems to have been averted when Simon comes up with a plan to lower Joe down 300 feet at a time, using the longest rope they have. Unfortunately, this turns out to be a classic **false dawn**. Simon, unable to see or hear Joe at the bottom of the rope, leaves him hanging over a cliff. With the rope fully extended, there's nowhere for Joe to go. Simon debates the options: wait for Joe to slowly drag them both to their deaths, or cut the rope. He cuts the rope.

HOLE

Cut to the B story. Richard, waking up the next morning, is certain that both of his friends must have died. Then Simon turns up with the grim news. He'd tried searching and yelling for Joe as best he could – without falling into the crevasse in which Joe had landed – but to no avail.

But for Joe, who had in fact survived the fall, it's a three-day-long **dark night of the soul**. Can you imagine: crawling and hopping for miles with a broken leg? Every hop, Joe explained afterwards, he would almost faint with the pain. And remember, even setting aside the broken leg, Joe had just completed the climb of his life, and was suffering from frostbite and extreme dehydration. The trial reaches its nadir with a surprisingly comic turn. Joe, who has by now accepted his inevitable death, hallucinates that he's lying beaten-up in a pub car park, while his brain plays Boney M's novelty hit 'Brown Girl in the Ring' on an incessant loop. In true 'darkest before the dawn' fashion, though, it's this lowest ebb that gives Joe the strength he needs for a final push. He doesn't expect to survive, but he doesn't want to die to Boney M. *Making preparations*, Joe steels himself for one final push.

And soils himself. Or so he thought. This was a regular occurrence during his ordeal. But through the haze of dehydration and agony, Joe somehow realizes that the urine and faeces he can smell are not his own. Or not just his own. He realizes he's crawling through the campsite latrines. The tents are within touching distance. He's *executing the plan*. He's going to make it.

Inevitably, though, reality has a *high tower surprise* in store. When Joe shouts, there's no reply. Of course – they must have left. Why on earth would Richard and Simon have hung around, knowing full well that Joe must be dead?

It's time for our hero to *dig deep down*. But where does the solution come from in the Hole masterplot recipe? That's right – from the B story; from some seemingly irrelevant third character. Let's hear it in Richard's own, very poignant words:

> I woke up, not knowing why. And was aware of this kind of strange atmosphere. I could hear the wind howling outside the tent. And I started hearing something. It did slowly dawn on me that really the only thing it could be, would be Joe outside shouting. But that was completely impossible, because he was dead.[4]

Quickly, our heroes execute a new plan: getting Joe onto a donkey and riding 200 agonizing kilometres to the nearest hospital in the capital, Lima. Joe survived, and his account of the ordeal, *Touching the Void*, went on to become both a bestselling book and an award-winning movie.

The extent to which his ordeal follows the Hole masterplot recipe would not have been lost on Joe Simpson, a hugely talented writer who had studied theatre at the University of Edinburgh. Indeed, as he later told *The Arts Desk*, it was a similarly grim real-life Hole story, *The White Spider*, that inspired him to start climbing in the first place.[5] What is more, he clearly understands why his story resonates with so many people. 'People think, I'm going to get into deep shit someday. I'm going to get cancer, I'm going to lose my job – whatever,' he told *The Guardian* in 2022. 'What would I do in that situation?'[6]

In the end, narrative became Joe's very reason for living. He kept going, not because he expected to survive – he didn't – he just wanted the story to have an ending: 'I thought that if I crawled down to the riverbed, someone would definitely find my body. I wasn't expecting to meet anybody but just to crawl to the end of the end game, to die there'.[7]

Such is the power of narrative: that even after giving up on life itself, we still have the will to see a story through to the very end.

THE SCIENCE BEHIND THE STORY

Masterplot recipes have long been an art form; but – like everything else these days – they are being turned into a science via the application of artificial intelligence (AI). Apple researcher Marco Del Vecchio used an AI to automatically analyse the story arcs of over 6,000 movies (well, their subtitles: AIs aren't quite clever enough – for now – to understand what they see). Having assigned each movie to a masterplot recipe, Del Vecchio and his team then looked to see which recipes are most commonly used, and which earn the most at the box office.[8] It turned out that Hole movies took both crowns. Almost 1,600 of the 6,000 movies analysed used the Hole recipe, as opposed to less than 1,000 for most of the others (Underdog was second with around 1,400). In terms of box office, Hole movies – with an average gross domestic (US) revenue of $37 million per movie – out-earned every other type. The other types didn't differ significantly from one another – an Underdog movie, a Sacrifice movie and a Quest movie will – all other things being equal – do similar box office. But a Hole movie will – again, all other things being equal – earn more; on average, about $6 million more. And it's a similar story for novels (at least, if we restrict ourselves to e-book downloads, which are easier to analyse using automated tools).[9]

This raises the question of just why the Hole masterplot recipe is more lucrative than the other types. Are Hole movies critically acclaimed? Are you kidding? Whether looking at critics' scores or IMDb's user ratings, Del Vecchio and colleagues found that Hole movies showed the lowest scores of all. Neither were they likely to win Oscars or other awards (Underdog movies swept the board). But while Hole movies have the *lowest* ratings, they also have the *most* ratings. The average Hole movie is reviewed by 133 critics and 78,000 IMDb users (as opposed to, for example, 117 critics and 67,000 IMDb users for Icarus movies). This gives us an important clue as to why Hole movies generate good box office: they generate buzz! When a movie that follows the Hole recipe comes out, people talk about it. Often, the things they're saying about it aren't complimentary, but that doesn't matter. As they say, there's no such thing as bad publicity (unless it's an obituary notice).

UNDER THE INFLUENCE

Why do Hole movies generate the most buzz? An intuitive answer is that this masterplot affects viewers – perhaps more than any other – by presenting dramatic situations and inviting them to think about – and *talk* about – what they'd do if they ever found themselves in this situation. There's a bomb on the bus! A meteor headed right for Earth! Snakes on the plane! A Terminator from the future out to kill you! What would you do? The fact that these situations are extremely unlikely isn't a problem. In fact, it's the whole point: nobody makes a movie that asks what you'd do if there *wasn't* a bomb on the bus, and you enjoyed a relaxing and uneventful journey.

This intuitive answer, it turns out, is backed by science. Neuroscientists have found important overlaps in the brain activity that we exhibit when doing something ourselves and simply watching others do the same thing. If I were writing this book thirty years ago, I would have told you that this phenomenon is all down to 'mirror neurons'. In 1992, a team of neuroscientists from Parma in Italy identified, in macaque monkeys, brain cells that fire both when a monkey grabs a piece of food and when they watch a different monkey – or even a human experimenter – performing the same action.[10] Similar findings were reported in brain-scanning studies with humans, even when the actions were shown in video recordings, rather than acted-out in real life.[11] These cells were quickly given the evocative name of 'mirror neurons', and generated a great deal of excitement in the scientific literature. For example, in 2000, the neuroscientist V. S. Ramachandran called mirror neurons 'the driving force behind the "great leap forward" in human evolution'.[12]

What got neuroscientists so excited was the idea that when we see someone – for example – wrestling with the steering wheel to avoid a head-on collision, we're not just passively observing, but actively simulating the scenario for ourselves. Although we might not actually move our hands, our minds are silently pantomiming the patterns of brain-cell firing that would be happening if we were. We're understanding the action 'from the inside'; what it feels like to be carrying it out. The same is true, so the mirror-neurons story goes, for emotions. We recognize the facial expression of, say, fear by activating the same brain areas that would be active if we were experiencing fear first-hand. Vittorio Gallese, one of the co-discoverers of mirror neurons, lost no time in spelling out the implications for film viewers. In his book *The Empathic Screen: Cinema and Neuroscience*, Gallese wrote that mirror neurons underlie not only 'our ability to share attitudes, sensations, and emotions with the actors' but even 'with the mechanical movements of a camera simulating a human presence'.[13]

It sounds almost too good to be true, and – alas – it probably is. Although the scientific debate continues, most neutral observers are these days rather sceptical. As long ago as 2014, another neuroscientist, Gregory Hickok, published a book titled *The Myth of Mirror Neurons*, spelling out the case for the prosecution. The details are technical, but essentially his argument is that the advocates of mirror neurons have gone far beyond what the data actually show. Nobody disputes the existence of brain cells that fire both when we perform a particular movement and see others doing so; it's just that this doesn't necessarily mean that we're 'simulating' the movement; perhaps these cells are simply involved in both perception and action.[14]

That said, although the mirror-neurons story is likely to be a grossly oversimplified one, nobody doubts that there is at least some similarity in the brain responses that we show when viewing an action or emotion and experiencing it first-hand (there can't be *too much* overlap, or we'd be constantly misinterpreting others' actions as our own!). And, fascinatingly, this overlap does seem to reflect what we think of as empathy. People who score high on empathy questionnaires do show greater 'simulation' of others' emotions in brain-scanning studies.[15] Similarly, those who are better at identifying sadness or anxiety in others seem to be more capable of experiencing these emotions second-hand. For example, when you're anxious, your heart rate speeds up and you start to sweat. Super empathizers – those who are great at picking up on anxiety in others – show big increases in heart rate and sweating when others show signs of anxiety.

But apart from literal psychopaths who, interestingly, don't show these effects at all, the same is true to some extent for all of us.[16] In fact, I'm sure you don't need me to tell you this: no doubt you've felt your pulse race and your skin dampen in the air-conditioned cool of the cinema as the on-screen hero finds themself in a life-threatening situation. What the science tells us, then, is that our intuitions are spot on: the reason we find Hole stories uniquely gripping – even as they are panned by the critics – is that we really do, *literally*, experience these terrifying yet exhilarating emotions vicariously, but from the comfort of our padded cinema seats.

PLOT TWISTED

As we've seen, Hole stories are uniquely powerful. Stories that follow this masterplot recipe generate more buzz and more box office than others, because we experience – literally – the rollercoaster of emotions that our hero is going through. It's no surprise, then, that the Hole masterplot recipe is particularly ripe for abuse. Politicians of all stripes know all too well that the way to galvanize their supporters is to promise to dig us all out of the deep, dark hole in which we find ourselves. And if there isn't a hole, no matter, just keep insisting that there is one *right there* – look, can't you see it?

The thing is, we humans are all too ready to believe that we're in a deep hole, even when the evidence suggests otherwise. This is neatly illustrated in a recent study by the psychologists Adam Mastroianni and Daniel Gilbert.[17] They trawled academic literature for surveys with questions like, 'Right now, do you think the state of moral values in this country as a whole is getting better or getting worse?' In a rather heroic effort, they managed to unearth surveys spanning more than twelve million people across sixty countries all around the world, dating back to 1949. I don't need to tell you the results, because I'm sure you have the same view yourself: the moral values of whichever country you currently find yourself in are undergoing an unprecedented decline. Although this view of moral decline is mostly strongly held by conservatives and older people, it's held to some extent by everyone, everywhere and everywhen.

But it's almost certainly a myth. As well as asking about *changes*, many of these surveys asked people directly about *current* levels of moral behaviour. 'Were you treated with respect all day yesterday?' they asked, or 'Within the past 12 months, have you been assaulted or mugged?' or 'During the past 12 months, have you let a stranger go ahead of you in line?' Mastroianni and Gilbert found that the responses that people gave to these questions were essentially unchanged since the middle of the last century. Things are not getting worse. So why do we all think that they are?

The authors identified two psychological quirks – though there are probably many more – that contribute to this illusion. The first is that people are more likely to seek out and pay attention to negative than positive information about others (a tendency that the media is only too happy to encourage). In support of this idea, Mastroianni and Gilbert found that when we're asked about people that we know well – rather than about society at large – we generally say that standards of morality have *improved*, rather than declined. The average best friend of today is – according to this study – *more* morally upstanding than the average best friend of the 1960s. It's only when we step outside of our immediate circle that we're on the lookout for bad news; probably because we're looking for evidence that we ourselves are better people than average (a phenomenon that we met in Monster).

The second quirk of our psychology that contributes to the myth of declining morality is that when we look back on the past, we do so with rose-tinted spectacles: bad events are more likely to be forgotten, or to be reinterpreted as not so bad, or even as good. In support of this idea, Mastroianni and Gilbert found that we report moral decline only within our own lifetimes. On average, a person born in 1980 thinks that moral standards were better in 1980 than in 2000, and worse still in 2020. But they don't think that moral standards were better in 1940 than 1960, or better in 1960 than 1980, since their rose-tinted spectacles don't cover the period before their birth. In short, we all seem to believe that the moment we were born, some 'other people' (not people that we know) started digging a hole, and it's been getting deeper ever since.

As a consequence, politicians and other cynical actors who say, 'You've never had it so bad' are preaching to the converted. We already think society is going to hell in a handcart, and are primed and ready to hear news that supports that view.

I'm sure you can pick your own example of a politician-manufactured crisis. We've already touched elsewhere in this book on Brexit. In 2015, the year before Britain's ultimately disastrous vote to leave the European Union, the issue was barely on the public's radar, with only 1 per cent citing it as the most crucial issue facing the country.[18] But various politicians found that the public were very receptive when they pointed out an EU-shaped hole in British society, and the rest is history.

In 2003, the USA led an invasion of Iraq, warning of its weapons of mass destruction programme and citing possible links to al-Qaeda and the 9/11 terrorist attacks. Although intelligence failure no doubt played some role, a 2019 fact-check by *The Washington Post* concluded that 'the Bush administration also chose to highlight aspects of the intelligence that helped make the administration's case, while playing down others'.[19] Certainly, no WMDs or al-Qaeda links were ever found.

In 2019, when continuing to make the case for his border wall with Mexico, Donald Trump blamed illegal immigrants for drug smuggling and increased crime. This despite the fact that the Drug Enforcement Administration had already noted that most drugs are brought in through legal points of entry (for example, mixed in with legal imports), while independent studies found *lower* rates of crime amongst illegal immigrants than US-born citizens.[20] As these examples show, Trump wasn't the first to mix up a poisonous concoction using the Hole recipe for political gain, and he certainly won't be the last.

It works.

HAPPY ENDINGS

It's rather ironic, then, that now we all find ourselves right at the bottom of a very deep and – for once – very genuine hole, almost all of us are doing almost nothing about it. I'm returning, of course, to the discussion of climate change that I began in the Sacrifice chapter. There I argued that we don't yet have our narrative right, a claim for which the widespread use of the term 'climate change' is further proof. This phrase is a euphemism for 'the impending extinction of the human species'. It is like calling the threat depicted in *Armageddon* or *Deep Impact* an 'asteroidal planetary adjustment'. So, can the Hole masterplot help us get our story straight and, like Bruce Willis or Robert Duvall, save the world? At this desperate point, anything is worth a try.

First, just how bad are things going to get? Well, I could reel off facts and figures; but in a book about the power of stories, that would be just perverse. The whole premise of this book is that we absorb, understand and process information only when it is skilfully mixed together following a masterplot recipe. If you will indulge me for a couple of pages, then, here is my story, *2101: A Spanish Odyssey*.

HOLE

A pitch-black screen. 'Barcelona, 2101' in white text at the bottom. As the camera pans right, we realize we are in a girl's bedroom. On a table illuminated by soft white LED night light, we see a framed photograph of a smiling middle-aged man in army fatigues. Partly obscured by the picture, we see a digital alarm clock: 23.58. Panning further right, we see a girl – seven years old, Mediterranean appearance – asleep under a thin white sheet. 'Sofia!' The girl's mother, Maria, rushes into the room stuffing bank notes into a dark green rucksack. She shakes the girl awake, grabs a pink rucksack from under the bed and stuffs it with a handful of clothes from Sofia's drawers, shouting 'Vamos, vamos!' She thrusts the pink rucksack into Sofia's arms and ushers her out the room. On the landing, a boy a couple of years younger is asking for his iPad. 'David!' shouts Sofia. 'Vamos! Vamos!!'

The family run out of the house, Maria carrying David with one arm, her car keys in the other. She looks at the drive, then spots the car floating at the other end of the street. She shoves the keys into her pocket and holds David with both arms. They wade through the streets with the water up to Sofia's middle. The street is hell, with everyone crying and yelling for missing relatives; the fast-flowing water is filled with cars, bikes, trash, tree branches, screaming people. Thunder and lightning in the sky, torrential rain pouring down.

As the family wade through the street, Maria gets out her phone and tries to call Ángel. 'I'm calling Daddy,' she tells the children. No signal. As she tries again, Sofia is hit by a floating branch and falls. Maria drops the phone and grabs her. She manages to pull her up but as she does so is hit by another branch, stumbles herself and loses her grip on David. He floats away in terror as they all scream. Fade to black.

Fade back in, to the same scene at first light, the sun looking incongruously beautiful on the horizon. Maria and her daughter are still searching and screaming for David. Sofia is struggling to keep her head above the water; her face is bloody and bruised. She is hit full in the face by a floating crate and passes out. As Sofia drifts in and out of consciousness, Maria continues to look for David, while carrying her daughter. But she can't do it; Sofia is too heavy to carry for more than a few seconds at a time. 'We've got to go,' she says. They set off, uphill.

An army tent in the desert, huge satellite dishes on poles. Ángel is half watching American TV with Spanish subtitles. It's a sixty-second summary of the day's big stories worldwide: wildfires in Los Angeles (the Hollywood sign ablaze); famine in the UK (shots of UN food parcels being handed out in Trafalgar Square); Visa, Mastercard and Apple Pay announce suspension of all services outside the US; pollution deaths in Delhi reach an estimated ten million; one-year anniversary of the European water war; major new floods in Tokyo, Miami and Barcelona. There is no footage, but the affected areas are shown on a map. Ángel exchanges glances with the other soldiers. A few slip him quarter-full water bottles and half-eaten energy bars as he surreptitiously packs his bag.

Maria and Sofia have reached Tibidabo, an abandoned amusement park at the top of a mountain overlooking the now-underwater city. The sun is high in the sky, and people are clearly suffering from the heat, but so many are packed into the park, only a few have room to sit down. The handful of shady spots are taken by elderly people lying flat out, many already dead of heatstroke. People are arguing, fighting and bartering agitatedly over food and water.

A cheer goes up from the crowd. A group of police officers have broken into a kiosk. While the others patrol the crowd with their handguns, two officers hand out ancient-looking popcorn, one handful per person, and water:

the soda gun is still working, and they pass out cardboard Coke cups with an inch of water in each. A light rain starts to fall, though the sun is still shining. People hold out their tongues, their empty Coke cups, their shoes. The mood in the crowd visibly lifts a little.

But as night falls, things turn even uglier than before, with full-on fist fights breaking out everywhere over food and water. The handful of police try to keep order, but are quickly overpowered and their guns taken. Fade to black over screams, gunshots, sounds of general chaos.

First light. Maria and Sofia wake up in a forest behind the amusement park. The contents of their bags are strewn across the ground – all the bottles and food wrappers are empty. Dead birds are everywhere. Maria takes a stick and tries to dig for water, but the soil is dry as far down as she can dig. She finds a worm, eats half and offers half to Sofia, who shakes her head, crying.

Ángel has arrived in the outskirts of Barcelona and finds an abandoned petrol station on the motorway, its sign showing €10 per litre. It looks relatively intact, but when he goes inside, he finds it's already been cleaned out. Heading back outside, he tries the petrol

HOLE

pump in frustration. Petrol shoots out as the dial spins rapidly. A thought occurs to him; he tries the car 'air and water' machine. Nothing. He tries the water tank in the car wash. Nothing. But in the office of the car wash, he smashes open the bottom desk drawer with his gun. Jackpot: half a bottle of Coke, some paprika-flavour crisps and a dried-out orange.

Heading outside, he sees the mountaintop theme park in the distance and heads there. It's dusk as he arrives and searches frantically for his family. He tries to ask around but everyone he speaks to is dead, dying or incoherent. Eventually he finds Maria and Sofia, but it is clear they have been dead for some time. Screaming and crying, he stamps on the orange and fills his mouth with crisps and Coke. Close up of Ángel's Adam's apple as he swallows then opens his mouth. Gunshot. Close up of his Apple watch as he lies dead on the floor. But all it shows is the red 'needs charging' symbol.

Film rights are available via my agent.

OK, so this rather heavy-handed tale is not going to win any Oscars. But it does, I hope, illustrate – in a way that mere facts and figures cannot – what the world will be like in 2101, assuming a middle-of-the-road estimate. That's right: this isn't some outlandish worst-case nightmare. My story is based on the

IPCC's median prediction of a rise in the global average temperature of around 4°C by the end of this century. Yes, despite the ironically futuristic-sounding title, 2101 isn't some unimaginably distant date. A real-world Sofia would be the granddaughter of a child being born right about now. And although the individual plot points are, of course, fictional, they are based on genuine expert predictions, as summarized in David Wallace-Wells's book *The Uninhabitable Earth*. A rise of around four degrees really would see direct heat death in places like Spain (in hotter places like India, we will see direct heat death in even a *best-case* scenario of a two-degree increase). Death by drowning will be commonplace. Every beach in the world – not just Barcelona's – will be underwater, as will Facebook's current headquarters, the Kennedy Space Center, the White House and (every cloud has a silver lining) Donald Trump's Mar-a-Lago resort. Fires in Los Angeles are already common, with more than eight million acres of California having already burned.[21] Water shortages are already inevitable. Freshwater lakes have shrunk or disappeared, with – for example – the former fourth largest lake in the world, the Aral Sea, having already lost 90 per cent of its water. These shortages will lead not only to death by dehydration, but to widespread starvation, with 70 per cent of the world's fresh water used not for drinking but agriculture. As a result, it is almost impossible to see us avoiding water wars or the societal collapse that will be triggered when people cannot access food or water.

But we are in danger of slipping back into facts and figures. Don't think of the millions who will die. Think of just one person: of 'Sofia' or 'David'. If you've recently had a child, death by climate change is the *most likely* – not just 'possible' – fate of their grandchildren.

Unless.

Unless we turn things around. So, how have we been doing?

In 2015, at the UN Climate Change Conference in Paris (COP21), 193 countries plus the European Union agreed a target of limiting the global temperature increase to 1.5°C above pre-industrial levels. This 1.5-degree figure was chosen not because this will avert disaster entirely – we are already experiencing severe negative impacts at the current 1.1-degree increase – but because it's just about the best we can hope for at this stage. And, in fact, the agreement is only to limit the increase to a fairly disastrous 2 degrees (direct heat death in many parts of the world, remember), while 'pursuing efforts' to limit the increase to 1.5 degrees. To hit this 1.5-degree target, we need to reduce global emissions by 45 per cent by 2030, and to reduce them to zero by 2050.

First, the bad news: we are not on track to reduce global emissions by 45 per cent by 2030; not by a long chalk. In late 2022, the Bezos Earth Fund, Climate Action Tracker, ClimateWorks Foundation, the United Nations High-Level Climate Change Champions and World Resources Institute published a joint report on progress so far.[22] It looked at forty different indicators of progress, from reforestation and green electricity generation to the share of kilometres travelled by passenger cars versus public transport. For *not one* of these indicators was progress happening at a sufficient rate to meet the 2030 targets. For only *six* are we 'heading in the right direction at a promising but insufficient pace'.

The UN's climate-change report painted a similarly bleak picture. Far from limiting global warming to 1.5 degrees, or even 2 degrees, we are on course for a 2.8-degree increase by 2100 on the basis of governments' current policies. Perhaps even more damningly, the report found that even if countries stick to all of the pledges they made at the 2021 COP26 conference in

Glasgow – a pretty big 'if' – we will still be on course for an increase of 2.4–2.6 degrees. Indeed, on a global level, we are not yet cutting emissions *at all*. Far from cutting emissions at rates of around 5 per cent a year, which would be needed to meet the 2050 target of net zero, we are still increasing emissions, at a rate of around 1–2 per cent a year.[23] A 2022 study by the World Meteorological Association found that, for both methane and carbon dioxide, rates of emissions are accelerating rather than decelerating.[24]

But there is some good news.

First, according to an analysis from the International Energy Agency,[25] the use of fossil fuels (gas and oil) is expected to peak in 2025 and then drop off. Why? Not because of countries' or companies' benevolence, but through their sheer self-interest. Russia's 2022 invasion of Ukraine sent already-high gas and oil prices into the stratosphere, meaning that governments are turning to renewables – wind, water and solar power – as the cheaper option.

Second, a clear solution is already on the table. In 2022, Saul Griffith, winner of a prestigious MacArthur 'Genius' grant, published a booked called *Electrify*.[26] Griffith argues that if we convert absolutely everything that we humans use to run off electricity rather than fossil fuels, we can meet the COP targets, without having to reduce our consumption. Best of all, we already have all of the relevant technology – electric vehicles, electric heat-pumps, electric cooking – so saving the planet is just a matter of rolling them out on a global scale and generating all of the necessary electricity. But *could* we generate all of the necessary electricity without burning fossil fuels? According to Griffith, yes: if we continue to add wind, water, solar and nuclear capacity 'at the rate that we have . . . over the past few decades', we will be able to stop burning fossil fuels entirely by 2037.

HOLE

The hard part, of course, is getting any of this to actually happen. Can the Hole masterplot recipe give us any clues about the best strategy here? I believe it can. After all, 'EVERYONE IN THE WORLD IS GOING TO DIE' is not only one of the most familiar tropes in Hollywood, but one of the most lucrative at the box office: *The Day After Tomorrow*, *Contagion*, *Outbreak*, *Pandemic*, *Train to Busan*, *Meteor*, *Armageddon*, *Deep Impact*; the titles speak for themselves. So how do the Hollywood heroes save the world?

First, it's useful to think about what doesn't happen in the Hollywood version. In *Meteor*, *Armageddon* and *Deep Impact* (all three are more or less the same movie), Earth is threatened with an 'extinction-level event' in the form of a giant comet. What is humanity's response? Do we say, 'Look, you can't expect the government to solve all your problems – What are YOU doing about the comet problem?' (the Sacrifice framing that I rejected in the previous chapter). Do we say, 'Gee, it would be nice to stop the comet, but it'd be awfully expensive. And what about the people who the comet *isn't* going to hit? Can we really expect them to foot the bill?' Do we say, 'Oh well, being hit with the comet is inevitable; the question now is how we transition to our post-comet future?'

Of course we don't; these options are clearly ridiculous. But these are essentially the solutions that we've pursued so far with our real-life extinction-level event. Indeed, climate-change experts are increasingly singling out the 'What are YOU doing?' Sacrifice approach as one of the biggest impediments to solving climate change. In *The Uninhabitable Earth*, David Wallace-Wells talks about the phenomenon of 'plastic panic', which he calls a 'climate red herring'. One of the most high-profile environmental campaigns of recent years has been to reduce plastic waste, to the point where most fast-food restaurants offer only paper straws (to poke through the plastic cup lids!). Many governments have also taken action on microplastics – tiny pieces of plastic that find their way into the food chain. Of course, it would be nice to reduce plastic pollution, but placing this goal front and centre of the environmental movement is the quintessential case of rearranging the deckchairs on the *Titanic* (or throwing stones at the approaching comet). Production of new plastics accounts for just 4 per cent of global oil and gas use,[27] and some microplastics in the atmosphere might actually *cool* the planet slightly by reflecting sunlight back into space.[28]

It's a similar story with recycling, which, according to Project Drawdown[29] – a non-profit that calculates the impact of possible solutions – ranks in the bottom half of the effectiveness league table, below things like increased wind and electric power (the table toppers), plant-rich diets, forest restoration and managing methane leaks. Of course, recycling is better than not recycling. But the very real danger is people thinking that if we all just do a bit more recycling and use a bit less plastic, things will be OK.[30] They will not.

So, who does save the Earth in *Armageddon*, *Deep Impact* and all the rest? Not corporations, not ordinary citizens, but the government (usually, as the world's most powerful, the US Government). And what does the government do in the movies? Does it offer tax breaks to companies who promise to invest in finding a solution to the comet problem? Does it run awareness campaigns encouraging individuals to Sacrifice, to do their bit?

No. It buys Bruce Willis a new vest and sends him off to destroy the comet with a shit-ton of nuclear missiles. Pardon my language, but if we're going to avoid the situation where *everyone on the planet is dead*, governments need to stop fucking around.

What the Hollywood Hole masterplot recipe teaches us, then,

is this: what we need to do is hire a charismatic hero (it might well be the Terminator himself, given Governor Schwarzenegger's strong record on environmental issues), give him a briefcase stuffed full of billion-dollar bills (literally, since optics are everything), and set him the mission of buying out every gas and oil company in the world, and converting them into suppliers of green energy. And, given the unique power of stories, we should make a reality TV series documenting his efforts. *Hostile Takeover*, I'd call it, and I'd distribute it for free to every TV company and streaming service in the world. Let's see which big-oil CEO is keen to cast himself as the villain of humanity when Governor Schwarzenegger makes him a more-than-generous offer.

You may laugh, but I'm not joking. A very similar solution – albeit without the Arnold Schwarzenegger part – is being proposed by *Electrify* author Saul Griffith. Rather than buying out the fossil-fuel companies and forcing them to transition to green energy, Griffith argues that governments could simply purchase the companies' land, and all the gas and oil under it, leaving them dripping with money to build green-energy infrastructure. Either way is good! But whatever the precise details of the plan, it absolutely *has* to be framed as a positive, upbeat Hole story starring a charismatic hero, rather than as a dreary politician-led appeal for personal Sacrifice, accompanied by dryly apocalyptic facts and figures.

Don't just take my word for it. A recent study led by researchers at the University of Birmingham set out specifically to determine exactly which types of framings garnered most support for climate policies, across citizens of China, Germany, India, the UK and the USA (a total of 7,500 participants).[31] The climate messages that were shown to the participants varied in four different ways. First, the messages were given either a positive (opportunity) framing ('We can save the world if we do this') or a negative (threat) framing ('We'll all die if we don't do this'). Second, the messages were pitched at four different levels: what *you* can do (individual), what your local community can do (community), what your country can do (national) or what we, globally, can do (global). Third, the messages mentioned different timescales: what we need to do right now, by 2030 or by 2050. Finally, the messages were focused on different consequences of climate change: the economy, migration, health (i.e. impacts on people) and the environment.

Before getting into the details of the findings, let's consider what most current messaging looks like. As we'd expect, given the current dominance of the Sacrifice narrative, it tends to have a negative framing, to focus on individuals, and on the COP goals for 2030 and 2050; for example, 'If we don't all reduce our consumption, the planet will see three degrees of warming by 2030, and we will miss our 2050 goal of net zero.' But according to the results of the study, this is almost exactly the wrong approach. The most effective messages had a positive framing, a global focus and an immediate time frame, as well as being geared around health or the environment. In other words, they adopted the Arnold Schwarzenegger Hole framing: 'If the *governments of the world* take this joint action *right now*, we will all live *long and heathy lives* well into the future, and have a *pleasant environment* to enjoy.'

If we heed the lessons of this study, and put a hero-led global plan into action right now, then it's just possible that the Hole masterplot recipe could save the world.

10.

THE STORIES OF YOUR LIFE

What's the greatest story ever told?

When I asked you this question at the start of the book, I gave you the BBC's answer, from a poll of literary critics: *The Odyssey*. For many people, though – particularly people from my generation – the answer is the *Star Wars* saga.

Before we set off for a galaxy far, far away, though, I need to come clean about a little secret I've been keeping from you. Throughout this book, I've been writing as if each series of real-life events conforms neatly to one – and only one – masterplot recipe. In many cases, this is true. For example, Karl Bushby's attempt to walk around the world follows the Quest recipe to the letter, but has none of the ingredients of a Feud, Hole or Icarus. Leicester City's Premier League win is an archetypal Underdog story, and it would be all but impossible to interpret it as a Monster, Untangled or Sacrifice story. Often, though, a series of real-life events can be interpreted through any one of several different masterplots. Sometimes, it's a matter of perspective: one man's Underdog is another man's Monster. Sometimes, it's a matter of time frame. Nothing lasts forever, and if you hang around long enough after the triumphant climax of an Underdog, Quest or Hole story it will often turn into an Icarus.

THE STORIES OF YOUR LIFE

Nobody appreciates the flexibility of masterplots more than *Star Wars'* creator George Lucas. Indeed, by flipping perspectives and zooming in and out on particular characters and time frames, Lucas pulls off the masterstroke – across the *Star Wars* saga – of following each and every one of the masterplot recipes at once.*

The first movie in release terms, originally called simply *Star Wars*, perfectly follows the Quest masterplot recipe (Chapter 2). Luke Skywalker, a humble moisture-farm worker on the desert planet of Tatooine, gets a **call to action** when – while cleaning a droid, R2-D2 – he stumbles across a hologram SOS from Princess Leia, 'Help me, Obi-Wan Kenobi. You're my only hope.' His quest to save the princess takes in a **monster** (the Trash Squid, more accurately known as Omi the Dianoga), the **supernatural** (as he masters the Force), **travelling companions** (R2-D2 and C-3PO, Han Solo and Chewbacca) and **local helpers** (Obi-Wan 'Ben' Kenobi). For Luke, who has never previously left his home planet, **unworldliness** is everywhere, and he triumphs in his **final ordeal** – destroying Darth Vader's planet-sized weapon, the Death Star – thus setting up his **life-renewing goal**: training to become a Jedi Knight (a Quest that forms the backbone of the second film, *The Empire Strikes Back*).

There's even time for a mini, but perfectly formed, Hole story (Chapter 9). Fleeing the Stormtroopers on the Death Star, Luke,

---

* How many films? It's complicated. The examples in this chapter are drawn from the original trilogy (*Episodes IV–VI*; 1977–1983) and the much-maligned (in retrospect, *somewhat* unfairly) prequel trilogy (*Episodes I–III*; 1999–2005); though what has come to be called the main 'Skywalker saga' also includes the sequel trilogy (*Episodes VII–IX*; 2015–2019). Also 'canon' (as the fans say) are two standalone live action films, *Rogue One* (2016) and *Solo* (2018), an animated film (*The Clone Wars*, 2008), recent TV series (*The Mandalorian*, *The Bad Batch*, *The Book of Boba Fett*, *Obi-Wan Kenobi* and *Andor*) and – in general – any books, comics or video games released since September 2014.

Han, Leia and Chewie dive down a garbage chute and **into the hole,** in this case a trash compactor. There's a **false dawn** when they manage to subdue Omi, who is living in the trash, before the walls start closing in, literally. There's a **dark night of the soul** – 'We're all going to be a lot thinner' – and our heroes **execute their plan** of bracing the walls of the compactor with a metal pole but – in a **high-tower surprise** – it doesn't work. **Digging deep down,** they find a solution from the B story, in the form of R2-D2, who has been winding up the guards and generally mucking about upstairs. They execute the new plan – radioing the droid to shut off the compactor – climb **out of the hole** and survive.

While all this is going on, an Untangled story (Chapter 3) is playing out in the background. In a story that was mirrored in real life by the actors concerned, uptight Luke and carefree Han battle for the affections of Princess Leia in a **bizarre love triangle.** There's an **initial state of chaos** as the first movie (*Star Wars*) hints at a burgeoning relationship between Luke and Leia – although it goes no further than a peck on the cheek at this stage – while Han and Leia dislike each other intensely. But in *The Return of the Jedi* (the final movie of the original trilogy), we learn – **concealed or mistaken identity** alert – that Luke and Leia are brother and sister. They even have a bit of a smooch (before they find out that they're siblings, I hasten to add), but this is really just a ploy on Leia's part to make Han jealous. Because, yes, the **untangling,** when it arrives, is the inevitable triumph of laid-back Han over uncool and brooding Luke, as Han and Leia are married.

If we turn our focus away from the heroes and to the Dark Side, we can see a classic Icarus story (Chapter 4). Like Luke's Quest, this Icarus story plays out across the whole nine-film trilogy of trilogies (original, prequel, sequel), but is thrown into sharpest relief in the final film of the prequel trilogy, *Episode III – Revenge of the Sith*. In the first two prequels, we saw cute little Anakin Skywalker rise from Tatooine slave to trusted Jedi – but in the third, **dissatisfaction** rears its head when the now grown-up Anakin is troubled by recurring visions of his wife, Padmé, dying in childbirth. Anakin's temptation comes when Chancellor Palpatine – up until now, posing as one of the goodies – promises that if Anakin comes over to the Dark Side of the Force, he will master the powers needed to save his wife. This is clearly a **transgression** and places Anakin in a **dilemma**. He loves Padmé, but wants to stay loyal to the Jedi and, indeed, initially reports Palpatine to the Jedi Council. But Palpatine's proposal targets both of Anakin's **Achilles heels**: fear of separation from his loved ones (Anakin having left his mother behind on Tatooine), and his egotistical ambition (embodied in his lust for power, and the resultant grudge he holds against the Jedi for declining to promote him to the rank of Jedi Master). Anakin defects to the Dark Side and – **elation** – Palpatine anoints him as Darth Vader. But now **insatiability** rears its ugly head: Anakin never feels secure in his position or gets over his resentment of the Jedi, and plays a particularly gruesome part in Palpatine's Jedi massacre, slaughtering the younglings – child Jedi in training – at their temple. Padmé gives birth to twins (Leia and Luke), but it's too late. Anakin has sown the seeds of his own destruction by choking Padmé in a fit of rage and being so generally beastly that she loses the will to live. Anakin's greatest fear has come to pass and – in true Icarus fashion – it's all his fault. His **destruction** is complete.

Every movie in the *Star Wars* saga has its own Monster narrative (Chapter 5). In the first film, *Star Wars* (later retrofitted as *Episode IV – A New Hope*), the **unlikely hero** Luke takes up a **magic weapon**, his lightsaber, to **save the world** – or, in fact, the universe – from a **monstrous monster**. Darth Vader, in his black cape, gloves and mask, is one of the most visually (and sonically) terrifying monsters in movie history. Unusually, Luke doesn't actually kill this particular monster – which would also have killed off the next two movies – but does the next best thing by destroying Vader's Death Star, the planet-sized (and planet-destroying) weapon that is menacing the galaxy, and enjoys **the hero's reward** of acclaim and status at a lavish ceremony. But Vader is a more complicated character than he at first appears, with a sympathetic back story and ending; and across the grand sweep of the whole multiple-movie narrative, it becomes clear that Palpatine – and the fascism that he espouses – is the real monster.

In fact, the relationship between Luke and Darth is arguably best understood through the lens of the Feud masterplot (Chapter 6). The two rivals are **evenly matched** – each the leading Jedi of his generation, favouring hand-to-hand combat with ancient lightsabers – but are **mirror images**: Luke is the light side – brave, kind-hearted, naive; Vader is the dark side – cowardly, cruel, cunning. One respect in which their relationship is more Feud than Monster is that neither ever tries to kill the other; instead, each tries to convert their rival to the opposite side of the Force. **Reconciliation and redemption** come at the end of the original trilogy (*Episode VI – Return of the Jedi*), when the dying Vader takes off his mask to look at his son. 'I've got to save you,' says Luke. 'You already have,' replies Vader, now redeemed as Anakin.

Fascinatingly, if we stop Anakin's story at the end of *Episode II – Attack of the Clones*, it is a straightforward Underdog tale (Chapter 7). In the first prequel (*Episode I – The Phantom Menace*), we **meet our hero as a young child**. Although he comes from the humblest of **humble beginnings** – both Anakin and his mother are slaves to a junk dealer – he has some ineffable essence of greatness; **a destiny to be realized.** When Anakin's blood is tested for midi-chlorians – the microscopic life form that is responsible for the Force – 'the reading's off the charts'. After beating an (extremely) **ugly sister,** named Sebulba, in a pod race, Anakin's potential becomes clear to Jedi Master Qui-Gon Jinn, who whisks him off to join the Jedi. By the end of *Episode II,* Anakin is fighting alongside Obi-Wan Kenobi and even Yoda himself, and is rewarded with Padmé's hand in marriage (albeit in secret – Jedi being monks and all).

But if we zoom out once more, and take in the whole nine-movie Skywalker Saga, Vader's story arc ultimately follows the Sacrifice masterplot (Chapter 8). In the final showdown between Palpatine, Vader and Luke (at the end of *Episode VI – Return of the Jedi*), Vader unexpectedly fights off Palpatine to save Luke, sustaining fatal injuries in the process. Vader is **prepared to give up something of great importance** – not just his life, but his devotion to the Dark Side – out of an **obligation** borne of family ties: 'Now go, my son . . . Tell your sister you were right.'

The *Star Wars* saga, then, illustrates the flexibility of masterplots. By zooming in on a small part of the overall story, or focusing on a particular pair of characters, or cutting off a story in mid flow and rolling the credits, or zooming out to take in a plot line that arcs across all nine movies, we can apply any one of the eight masterplots we've met in this book.

And it's just the same for real-world examples. Take, for example, the Covid-19 pandemic.

If we were making a feature film, the opening shots would show the world in the record-breaking hot summer of 2019: sunbathers sipping cocktails in Miami, street kids playing football on Copacabana Beach, a toddler licking an ice cream at London Zoo. A secondary character – perhaps a lab assistant in Wuhan – would say something like, 'I know life seems tough right now, but cheer up – sometimes you don't know how lucky you are until it's all taken away from you.'

But what is the closing shot? Cranes over all the world's major cities, building gleaming new office blocks; or toddlers dying in their mothers' arms in the slums of Rio? Is it a Hole story (Chapter 9), in which we climbed out of the hole in early 2022, as governments around the world lifted all remaining restrictions? Or is it an Icarus story (Chapter 4), in which we unleashed a new and deadly virus on the world, thought – in our hubris – that we had beaten it, and left it to spread unchecked throughout the world for the rest of human history?

Right at the start of the pandemic, with Covid still mainly localized to Wuhan, China, the dominant framing was that of a plot-twisted malignant Quest (Chapter 2). Covid was something that you had to travel to China to find and bring back; in doing so – as in the classic Quest masterplot – changing everything for you and your community.

When the virus spread worldwide, and most countries introduced lockdowns, the dominant masterplot framing became Sacrifice (Chapter 8). Children whose schools were closed became isolated from their friends, and had to rely on already stretched to the limit parents for their education. Essential workers such as doctors, emergency-service workers and supermarket staff literally risked their lives every day serving the rest of us.

Those for whom it was all too much found some comic relief in the Untangled stories (Chapter 3) that emerged over lockdown, such as when UK Health Secretary Matt Hancock broke social distancing rules (and that's putting it mildly) to carry on an affair with his aide. Meanwhile, conspiracy-theory plot-twisted versions of the Untangled recipe took hold. Was Covid leaked from a lab in China, perhaps deliberately? Was Bill Gates putting microchips in the vaccines to – I can't remember the details – force everyone to use Microsoft Teams instead of Zoom?

The high point of the pandemic – in a good way – came when the first vaccines were given, impressively within twelve months of the sequencing of the virus. For about five minutes, everyone in the world – or, at least, in the wealthy countries that could afford the vaccines – united behind a Monster narrative (Chapter 5) in which many scientists had proved themselves to be true heroes.

Indeed, the story of the development of the first Covid vaccine was a classic Underdog tale (Chapter 7). Katalin Karikó, born in Hungary to a butcher and a bookkeeper (who raised Katalin in a house with no running water) had managed to work her way up to an adjunct professor position at the University of Pennsylvania Medical School. She applied for grant after grant to study messenger ribonucleic acid (mRNA), which gives cells instructions for making proteins, but they were all rejected. Yet she clung on and, eventually, she hit upon a method of introducing mRNA into the body in a way that prevented it from being rejected by the immune system. It was this technology that would ultimately lay the foundations for the Pfizer and Moderna Covid vaccines.[1]

But then the Feud (Chapter 6) that had always been bubbling along under the surface, particularly in the USA, came to the fore. The vaccines, miracles though they were, gave only limited protection, for a limited time. Other precautions would be needed too: boosters, masks, pre-emptive testing, rules on the number of people who could socialize together. The battle lines were drawn, in a way that seemed quite arbitrary. It could have been political conservatives who called for protecting the old and vulnerable at all costs, while liberals railed against depriving people of their freedom. As it turned out, of course, it was the other way around. Either way, the feud soon took on a life of its own as a proxy battle between the right-minded and the left-minded politics, one which had little to do with public health and everything to do with the culture wars.

Now, which of the above masterplot framings – if any – chime with your experiences of the pandemic?

It's not that some of these framings are 'right' and others 'wrong'. The point, just as we saw for the fictional *Star Wars* universe, is that masterplots are *flexible*. By zooming in on an individual series of events, or focusing on a particular player, or cutting off the story at a particular point, or zooming out to see the big picture, we can apply any one of the eight masterplots we've met in this book to the pandemic.

It's just the same for you. Whether you're seeking to understand, or to manipulate, the behaviour of others, or of yourself, you have a choice: which narrative masterplot recipe will you use for your framing?

When I first came up with the idea for this book, and even when I sat down to start writing it, I still didn't appreciate the full power or – crucially – the flexibility of masterplots. In the main, I still thought of them as amusing curiosities. Matching up masterplots to books I read, films I watched or real-life events I lived through felt like solving a crossword or a sudoku: fun, intellectually stimulating, even broadly educational, but no more than that.

But, as I wrote, the more scientific papers I read, the more experts I interviewed, and the more real-life stories I encountered, the more I realized that masterplots have the power to shape our destiny, not just as individuals, but as a species. My hope for this book is literally that it will save the world: that it will somehow find its way into the hands of the world leaders – the politicians, the CEOs, the influencers – who have the power to do something about climate change. My hope is that, having understood the power of masterplots to influence our behaviour – and the power that we have to impose our chosen masterplots on real-world events – that they will finally act; that they will abandon the unprofitable Sacrifice framing and adopt the here-to-save-the-world Hole framing advocated by experts such as Saul Griffith. We can choose the masterplot we want the overarching story of humanity to follow. Please, let's not let it be Icarus.

If saving humanity is too ambitious a goal for this book, then here's a more modest one. Let's not let the bad guys win. Or, at least, not *every single time*, as they seem to be at the moment. Throughout this book, and throughout the last decade or so of human history, we've seen countless instances of bad actors twisting masterplots to their advantage. We've met megacorporations who portray themselves as Underdogs, gangs and families trapped in long-running Feuds, conspiracy theorists (Untangled), Eichmann and the Nazis (Monster), Donald Trump (both Underdog and Hole). The lesson of this book – particularly this final chapter – is that we don't have to accept these misleading framings. Masterplots don't (yet) belong to Google, Apple or Open AI. They belong to all of us, and we all have the option – the responsibility, even – to reject twisted masterplot framings and to advance universally beneficial ones.

And if even that goal is too ambitious, then here's an even more modest one. My hope for this book is that you, the reader, will be inspired to use masterplots not only to achieve your major life goals, but to find meaning in even the most unremarkable moments of your day-to-day life . . .

*

THE STORIES OF YOUR LIFE

It's Monday morning, and you're setting off for work. What are you doing, and why? Are you embarking on a Quest to – depending on your job – cure a disease (Monster), fight for the rights of the wronged (Underdog) or tackle climate change (Hole)? Or maybe you hate your job, and every working day is a Sacrifice borne out of your obligation to support your family. Perhaps Feud is your motivation. Maybe you're an academic who wants to stick it to the other side, a Nike employee who just hates Adidas or a defence lawyer who has a long-running enmity with a particular prosecutor. Maybe you're a CEO or investor who enjoys flying close to the sun with risky bets or, even better, watching your overstretched, over-ambitious rivals crash and burn (Icarus). Even if the job itself is meaningless, perhaps your office is an Untangled den of intrigue: who's having an affair with whom? And does *she* know? But didn't she catch *her* with *him* at the Christmas party?

Or maybe none of this applies, and you're just there for the money. OK, but – beyond the basics – why do you need money? To prove to the 'ugly sisters' who doubted you that, even though you didn't start out with much in life, you overcame the odds to triumph as the hero of your Underdog story? Because of some friendly competition – or bitter Feud – with your siblings over who has the most earning power? Maybe you're in the grip of the Monster of addiction, and need money to fund your treatment (or, for that matter, your habit). Maybe gambling is your vice and you're an insatiable Icarus who is constantly throwing good money after bad. Maybe you're saving up for a trip abroad: a highbrow Quest for knowledge in Athens, or a more worldly, Untangled-style quest to find the love of your life, or at least a holiday romance, in Mykonos. Or maybe you just want to keep a roof over your head and food on the table while you pursue your real passions outside of work: volunteering in a charity

THE STORIES OF YOUR LIFE

shop (Sacrifice), campaigning for political or environmental causes (Hole), following your football team (Feud) or forgetting about the serious world of work and enjoying spending time with your family, friends or hobbies (Untangled).

Whatever framing you choose, it's your life, and these are your stories to use as you will.

They are the Stories of Your Life.

NOTES

1. The Inside Story: Masterplot Recipes

1 While the three-act structure itself dates back to antiquity, its formalization as a model of how stories work is usually credited to Syd Field, and his surprisingly recent 1979 book: Field, S. (1979). *Screenplay: The foundations of screenwriting*. Delta.
2 Clark, A. (2013). Whatever next? Predictive brains, situated agents, and the future of cognitive science. *Behavioral and Brain Sciences, 36*(3), 181–204. doi:10.1017/S0140525X12000477
3 Bubic, A., Von Cramon, D. Y. & Schubotz, R. I. (2010). Prediction, cognition and the brain. *Frontiers in Human Neuroscience, 4*, 1094. https://doi.org/10.3389/fnhum.2010.00025
4 Zald, D. H. & Zatorre (2011). On music and reward. In J. Gottfried (ed.), *The neurobiology of sensation and reward*. Taylor & Francis.
5 Mandler, G. (1975). *Mind and emotion*. Wiley.
6 Leavitt, J. D. & Christenfeld, N. J. S. (2011). Story Spoilers Don't Spoil Stories. *Psychological Science, 22*(9), 1152–4. https://doi.org/10.1177/0956797611417007
7 Hollerman, J. R. & Schultz, W. (1998). Dopamine neurons report an error in the temporal prediction of reward during learning. *Nature Neuroscience, 1*(4), 304–9.
8 Demir, Ö. E. & Küntay, A. C. (2014). Narrative development. In P. J. Brooks and V. Kempe (eds), *Encyclopedia of Language Development* (pp. 393–7). Sage. https://doi.org/10.4135/9781483346441

2. Quest

1. Haynes, N. (2018, May 22). The greatest tale ever told? *BBC Culture.* https://www.bbc.com/culture/article/20180521-the-greatest-tale-ever-told
2. These ingredients and many of the fiction examples are taken and/or adapted from the chapters 'The Quest' and 'Voyage and Return' from Christopher Booker's *The Seven Basic Plots: Why We Tell Stories* (which treats the two as distinct masterplots). The same is true for Untangled (Booker's 'Comedy'), Icarus (Booker's 'Tragedy'), Monster (Booker's 'Overcoming the Monster') and Underdog (Booker's 'Rags to Riches'). Booker, S. (2004). *The seven basic plots: Why we tell stories.* Continuum.
3. https://www.westboundhorizons.com/situ
4. Siegel, S., Hinson, R. E., Krank, M. D. & McCully, J. (1982). Heroin 'overdose' death: contribution of drug-associated environmental cues. *Science,* 216(4544), 436–7.
5. The Charity Commission for England and Wales (2022, June 30). *Regulator announces statutory inquiry into The Captain Tom Foundation* [Press release]. https://www.gov.uk/government/news/regulator-announces-statutory-inquiry-into-the-captain-tom-foundation
6. Liu, Z. (2023, March 1). ChatGPT will command more than 30,000 Nvidia GPUs: Report. *Tom's Hardware.* https://www.tomshardware.com/news/chatgpt-nvidia-30000-gpus
7. Bennett, J. (2019, December 31). The top ten scientific discoveries of the decade. *Smithsonian Magazine.* https://www.smithsonianmag.com/science-nature/top-ten-scientific-discoveries-decade-180973873/
8. Substance Abuse and Mental Health Services Administration. Impact of the DSM-IV to DSM-5 Changes on the National Survey on Drug Use and Health [Internet]. Rockville (MD): Substance Abuse and Mental Health Services Administration (US); 2016 Jun. Table 3.20, *DSM-IV to DSM-5 Psychotic Disorders.* Available from: https://www.ncbi.nlm.nih.gov/books/NBK519704/table/ch3.t20/
9. Hartley, T. (2020, March 26). My psychosis. *Aeon.* https://aeon.co/essays/what-one-night-of-psychosis-felt-like-to-a-young-psychologist

NOTES

3. Untangled

1. Buckland, E. (2023, March 3). Shania Twain reveals ex-husband Robert 'Mutt' Lange is still with her former BFF 15 years after affair was exposed. *Daily Mail*. https://www.dailymail.co.uk/tvshowbiz/article-11818191/Shania-Twain-reveals-ex-husband-former-BFF-15-years-affair-exposed.html
2. Eames, T. (2023). The complicated history of how Shania Twain swapped husbands with best friend after ex cheated on her. *Smooth Radio*. https://www.smoothradio.com/news/music/shania-twain-husband-ex-best-friend-marriage/#:~:text=Shania%20met%20record%20producer%20Robert,in%20December%20of%20that%20year
3. The explanation given here is based on: Hurley, M. M., Dennett, D. C. & Adams, R. B. (2011). *Inside Jokes: Using Humor to reverse-engineer the mind*. MIT press.
4. The acid test for a theory of humour is whether it can be reverse-engineered to produce jokes that are at least somewhat funny. So, purely in the interests of science, you understand, I had a go. My 'process', if that's not too grandiose a term, was plucking words and phrases at random from the world around me: 'Word', 'Key performance indicators', 'light bulb', and trying to come up with alternative meanings that I could plausibly get a reader to commit to (e.g. 'Word' as an expression of agreement, rather than the software I'm using to write this).

> Speaker A: Microsoft sure do make some crappy software.
> Speaker B: Word.

> I keep locking myself out of the office. It's not going to look good on my key performance indicators.

> Q: How many men does it take to change a light bulb?
> A: One. You only need two if it's a heavy bulb.

> Q: How many men does it take to change a light bulb?
> A: Don't bother – you only need to change it when it's a dark bulb.

> What do you call the thing you sit on when you're at your desk?
> A chair!
> Ooh get you – you sound like the Queen when you sneeze.

Did you know that dropping just one letter from a word can be *fatal*?

That's preposterous! Give me an example.

I just did.

What did the sneaker say when he recognized his long-lost twin? Shoe!

I don't know why they call paying with your phone 'contactless' – I've got like 300 of the bastards in there.

But yeah, contactless payments have really taken over everywhere, haven't they? These days even the pickpockets have signs saying, 'We don't take cash'.

5   Cahn, L. (2023, August 22). 12 Conspiracy theories that actually turned out to be true. *Reader's Digest*. https://www.rd.com/list/conspiracy-theories-that-turned-out-to-be-true/
6   Lewandowsky, S., Oberauer, K. & Gignac, G. E. (2013). NASA Faked the moon landing – therefore, (climate) science is a hoax: An anatomy of the motivated rejection of science. *Psychological Science*, 24(5), 622–33. https://doi.org/10.1177/0956797612457686
7   Godwin, R. (2019, July 10). One giant . . . lie? Why so many people still think the moon landings were faked. *The Guardian*. https://www.theguardian.com/science/2019/jul/10/one-giant-lie-why-so-many-people-still-think-the-moon-landings-were-faked
8   Temperton, J. (2020, April 6). How the 5G coronavirus conspiracy theory tore through the internet. *Wired*. https://www.wired.co.uk/article/5g-coronavirus-conspiracy-theory
9   Pennycook, G., Cheyne, J. A., Barr, N., Koehler, D. J. & Fugelsang, J. A. (2015). On the reception and detection of pseudo-profound bullshit. *Judgment and Decision Making*, 10(6), 549–63. https://doi.org/10.1017/S1930297500006999.
    The statements can be found in the associated supplementary materials available at: https://sjdm.org/journal/15/15923a/supp.pdf
10 Pennycook, G. & Rand, D. G. (2020). Who falls for fake news? The roles of bullshit receptivity, overclaiming, familiarity, and analytic thinking. *Journal of Personality*, 88(2), 185–200. https://doi.org/10.1111/jopy.12476

NOTES

11 Ramsell, H. (2019, October 11). When I grow up I want to be . . . *Perkbox*. https://www.perkbox.com/uk/resources/blog/when-i-grow-up-i-want-to-be#:~:text=Those%20who%20don%27t%20 follow,unhappy%20in%20their%20dream%20rol

12 Headey, B. (2008). Life goals matter to happiness: A revision of set-point theory. *Social Indicators Research*, 86, 213–31. https://doi.org/10.1007/s11205-007-9138-y

13 Brooks, A. C. (2021, March 25). Are you dreaming too big? *The Atlantic*. https://www.theatlantic.com/family/archive/2021/03/how-follow-your-dreams-and-get-happier/618384/

4. Icarus

1 At the time of writing, electricity was £0.29/kWh in the UK and €0.112/kWh in Bulgaria. In Norway and Sweden, the rate was €0.051/kWh. Ofgem (2023, November 23). *Changes to energy price cap from 1 January 2024* [press release] https://www.ofgem.gov.uk/publications/changes-energy-price-cap-1-january-2024#:~:text=This%20will%20take%20the%20price,30%20 p%2Fday%20for%20gas. https://www.energyprices.eu

2 House of Commons Library (2024, March 22). *Rising cost of living in the UK*. https://commonslibrary.parliament.uk/research-briefings/cbp-9428/

3 (2016, October 16). The full text: Boris Johnson's secret article backing Britain in the EU. *Evening Standard*. https://www.standard.co.uk/news/politics/boris-johnsons-article-backing-britains-future-in-the-eu-a3370296.html

4 James, L. (2023, April 24). 'We've got no plan. What will we do?': Boris Johnson 'shock at Brexit result' revealed in new book. *Independent*. https://www.independent.co.uk/news/uk/politics/boris-johnson-brexit-no-plan-b2326008.html

5 Knock, E. S., Whittles, L. K., Lees, J. A., Perez-Guzman, P. N., Verity, R., FitzJohn, R. G., . . . & Baguelin, M. (2021, January 13). The 2020 SARS-CoV-2 epidemic in England: key epidemiological drivers and impact of interventions. MedRxiv. https://doi.org/10.25561/85146

6 https://www.theguardian.com/world/video/2020/mar/27/i-shook-hands-with-everybody-says-boris-johnson-weeks-before-coronavirus-diagnosis-video

7   Edwards, J. (2020, April 29). For years, Boris Johnson refused to say exactly how many children he has. *Business Insider.* https://www.businessinsider.com/boris-johnson-refuses-to-say-how-many-children-he-has-2019-11?r=US&IR=T
8   Watanabe, S., Weiner, D. S. & Laurent, S. M. (2022). Schadenfreude for undeserved misfortunes: The unexpected consequences of endorsing a strong belief in a just world. *Journal of Experimental Social Psychology, 101,* 104336. https://doi.org/10.1016/j.jesp.2022.104336
9   Singer, T., Seymour, B., O'Doherty, J. P., Stephan, K. E., Dolan, R. J. & Frith, C. D. (2006). Empathic neural responses are modulated by the perceived fairness of others. *Nature, 439*(7075), 466–9.
10 Huron, D. & Vuoskoski, J. K. (2020). On the enjoyment of sad music: pleasurable compassion theory and the role of trait empathy. *Frontiers in Psychology, 11,* 499421. https://doi.org/10.3389/fpsyg.2020.01060
11 Meeks, G. & Whittington, G. (2023). Death on the stock exchange: The fate of the 1948 population of large UK quoted companies, 1948–2018. *Business History, 65*(4), 679–98.
12 Sull, D. (1999, July–August). Why good companies go bad. *Harvard Business Review.* https://hbr.org/1999/07/why-good-companies-go-bad
13 Anonymous (2023, February). Miscarriage. *March of Dimes.* https://www.marchofdimes.org/find-support/topics/miscarriage-loss-grief/miscarriage
14 Anonymous (No date). Miscarriage: your questions answered. *NCT.* https://www.nct.org.uk/pregnancy/miscarriage/miscarriage-your-questions-answered#:~:text=Most%20miscarriages%20occur%20at%20the,first%2012%20weeks%20of%20pregnancy
15 Ludwig, L., Werner, D. & Lincoln, T. M. (2019). The relevance of cognitive emotion regulation to psychotic symptoms – a systematic review and meta-analysis. *Clinical Psychology Review, 72,* 101746. https://doi.org/10.1016/j.cpr.2019.101746
16 Duncan, C. & Cacciatore, J. (2015). A systematic review of the peer-reviewed literature on self-blame, guilt, and shame. *OMEGA-Journal of Death and Dying, 71*(4), 312–42. https://doi.org/10.1177/0030222815572604

NOTES

17 Alix, S., Cossette, L., Cyr, M., Frappier, J. Y., Caron, P. O. & Hébert, M. (2020). Self-blame, shame, avoidance, and suicidal ideation in sexually abused adolescent girls: a longitudinal study. *Journal of Child Sexual Abuse, 29*(4), 432–47. https://doi.org/10.1080/10538712.2019.1678543

18 Yusoff, M. S. B. (2015). A DEAL-based intervention for the reduction of depression, denial, self-blame and academic stress: A randomized controlled trial. *Journal of Taibah University Medical Sciences, 10*(1), 82–92. https://doi.org/10.1016/j.jtumed.2014.08.003

19 Gamble, J., Creedy, D., Moyle, W., Webster, J., McAllister, M. & Dickson, P. (2005). Effectiveness of a counseling intervention after a traumatic childbirth: a randomized controlled trial. *Birth, 32*(1), 11–19. https://doi.org/10.1111/j.0730-7659.2005.00340.x
As far as we can tell from the original research article, all participants identified as 'mothers' and 'women', the terms used in the original article.

20 Nikcevic, A. V., Tinkel, S. A., Kuczmierczyk, A. R. & Nicolaides, K. H. (1999). Investigation of the cause of miscarriage and its influence on women's psychological distress. *BJOG: An International Journal of Obstetrics & Gynaecology, 106*(8), 808–13. https://doi.org/10.1111/j.1471-0528.1999.tb08402.x

21 Again, as far as we can tell from the original research article, all participants identified as 'mothers' and 'women', the terms used in the original article.

22 Field, N. P. & Bonanno, G. A. (2001). The role of blame in adaptation in the first 5 years following the death of a spouse. *American Behavioral Scientist, 44*(5), 764–81. https://doi.org/10.1177/00027640121956485

5. Monster

1 Lewis, I. (2021, September 28). James Bond: Disability campaigners call for end to 'outdated trope' of villains with facial disfigurements. *Independent.* https://www.independent.co.uk/arts-entertainment/films/news/james-bond-disability-campaigners-villains-b1928355.html

2 https://www.youtube.com/watch?v=6EghiY_s2ts

3 Ward, J. H., Bejarano, W., Babor, T. F. & Allred, N. (2016). Re-introducing Bunky at 125: EM Jellinek's life and contributions

to alcohol studies. *Journal of Studies on Alcohol and Drugs, 77*(3), 375–83. https://doi.org/10.15288/jsad.2016.77.375

4   Zell, E., Strickhouser, J. E., Sedikides, C. & Alicke, M. D. (2020). The better-than-average effect in comparative self-evaluation: A comprehensive review and meta-analysis. *Psychological Bulletin, 146*(2), 118–49. https://doi.org/10.1037/bul0000218
    Storr, W. (2019). *The Science of Storytelling*. William Collins

5   Pronin, E. (2008). How we see ourselves and how we see others. *Science, 320*(5880), 1177–80. https://doi.org/10.1126/science.1154199

6   Leslie, I., (2017, October 13). The scientists persuading terrorists to spill their secrets. *The Guardian*. https://www.theguardian.com/news/2017/oct/13/the-scientists-persuading-terrorists-to-spill-their-secrets

7   Starr, D., (2013, December 1). The Interview. *The New Yorker*. https://www.newyorker.com/magazine/2013/12/09/the-interview-7

8   Leslie, I., (2017, October 13). The scientists persuading terrorists to spill their secrets. *The Guardian*.

9   Alison, L. J., Alison, E., Noone, G., Elntib, S. & Christiansen, P. (2013). Why tough tactics fail and rapport gets results: Observing Rapport-Based Interpersonal Techniques (ORBIT) to generate useful information from terrorists. *Psychology, Public Policy, and Law, 19*(4), 411–31. https://doi.org/10.1037/a0034564

10 Milgram, S. (1963). Behavioural Study of obedience. *The Journal of Abnormal and Social Psychology, 67*(4), 371–8. https://doi.org/10.1037/h0040525

11 Milgram (1963), p.371

12 Bilsky, L. Y. (2004). *Transformative justice: Israeli identity on trial*. University of Michigan Press.

13 White, T. (2018, April 23). What did Hannah Arendt really mean by the banality of evil? https://aeon.co/ideas/what-did-hannah-arendt-really-mean-by-the-banality-of-evil

14 Arendt, H. (1971). Thinking and moral considerations: A lecture. *Social Research, 38*(3), 417–46.

15 Dolan. E. W., (2019, November 17). Unpublished data from Stanley Milgram's experiments cast doubt on his claims about obedience. *PsyPost*. https://www.psypost.org/2019/11/unpublished-data-from-stanley-milgrams-experiments-casts-doubts-on-his-claims-about-obedience-54921

NOTES

Hollander, M. M. & Turowetz, J. (2017). Normalizing trust: Participants' immediately post-hoc explanations of behaviour in Milgram's 'obedience' experiments. *British Journal of Social Psychology*, 56(4), 655–74. https://doi.org/10.1111/bjso.12206
Perry, G., Brannigan, A., Wanner, R. A. & Stam, H. (2020). Credibility and incredulity in Milgram's obedience experiments: A reanalysis of an unpublished test. *Social Psychology Quarterly*, 83(1), 88–106. https://doi.org/10.1177/0190272519861

16 Caspar, E. A., Christensen, J. F., Cleeremans, A. & Haggard, P. (2016). Coercion changes the sense of agency in the human brain. *Current Biology*, 26(5), 585–92. http://dx.doi.org/10.1016/j.cub.2015.12.067

17 Milgram (1963), p.378

18 Much the same can be said about Zimbardo's equally famous prison study. Volunteers were randomly assigned the role of prisoner or guard in a mock prison (in the basement of the Stanford University Psychology Department) and the guards treated the prisoners in a thoroughly beastly fashion (though, again, reports get many of the details wrong, at least according to the French psychologist Thibault Le Texier:
Le Texier, T. (2019). Debunking the Stanford prison experiment. *American Psychologist*, 74(7), 823. https://doi.org/10.1037/amp0000401

19 Cesarani, D., (2007). *Becoming Eichmann: Rethinking the life, crimes, and trial of a 'Desk Murderer'*. Da Capo.

20 Shaw, J. (2019). *Making Evil: the science behind humanity's dark side*. Canongate.

21 Anonymous (2015, July 30). David Cameron criticised over migrant 'swarm' language. *BBC News*. https://www.bbc.co.uk/news/uk-politics-33716501

22 Pengelly, M. (2023, December 18). Trump's 'dehumanising and fascist rhetoric' denounced by top progressive. *The Guardian*. https://www.theguardian.com/us-news/2023/dec/18/trump-immigrants-rally-congress-reaction-pramila-jayapal

23 Helmore, E., (2023, December 17). 'He's dog-whistling': Trump denounced over anti-immigrant comment. *The Guardian*. https://www.theguardian.com/us-news/2023/dec/17/trump-denounced-anti-immigrant-comment

24 Hamilton, K. (2023, December 18). As border extremism goes mainstream, vigilante groups take a starring role. *Los Angeles Times*. https://www.latimes.com/world-nation/story/2023-12-18/arizona-border-militias-extremism-mainstream-immigration

6. Feud

1. In fact, *timshel* is itself a mistranslation and/or mistransliteration. Vowels are not normally written in Hebrew (since they can be inferred from the context), but the closest word *timshol* translates roughly as 'you will/shall rule' (over sin), meaning that the traditional translations have it right after all. See:
Levin, D. (2015). John Steinbeck and the Missing Kamatz in East of Eden: How Steinbeck Found a Hebrew Word but Muddled Some Vowels. *Steinbeck Review, 12*(2), 190–8. https://doi.org/10.5325/steinbeckreview.12.2.0190
2. https://en.wikipedia.org/wiki/Journal_of_a_Novel
3. https://en.wikipedia.org/wiki/One_red_paperclip
4. Grady, C., (2019, February 3). The Super Bowl halftime show controversies, explained. *Vox*. https://www.vox.com/culture/2019/2/1/18202128/super-bowl-2019-liii-53-halftime-show-controversy-maroon-5-travis-scott-big-boi
5. McClure, S. M., Li, J., Tomlin, D., Cypert, K. S., Montague, L. M. & Montague, P. R. (2004). Neural correlates of behavioral preference for culturally familiar drinks. *Neuron, 44*(2), 379–87.
6. Lau, R. R., Sigelman, L., Heldman, C. & Babbitt, P. (1999). The effects of negative political advertisements: A meta-analytic assessment. *American Political Science Review, 93*(4), 851–75. https://doi.org/10.2307/2586117
7. Nova, A. & David, J. E., (2018, November 3). The big soda companies are financing efforts to stop taxes on food and drinks: NY Times. *CNBC*. https://www.cnbc.com/2018/11/03/pepsico-and-coca-cola-fight-to-keep-sugary-drinks-from-being-taxed.html
8. McNamara, N. (2018, August 27). 'Yes! to affordable groceries' still backed by soft drink heavies. *Patch*. https://patch.com/washington/across-wa/yes-affordable-groceries-still-backed-soft-drink-heavies#
9. Tajfel, H. (2001). Experiments in intergroup discrimination. In

NOTES

M. A. Hogg & D. Abrams (eds), *Intergroup relations: Essential readings* (pp.178–87). Psychology Press.

10 Shariatmadari, D. (2018, April 16). A real-life Lord of the Flies: the troubling legacy of the Robbers Cave experiment. *The Guardian*. https://www.theguardian.com/science/2018/apr/16/a-real-life-lord-of-the-flies-the-troubling-legacy-of-the-robbers-cave-experiment

11 Hamlin, J. K., Mahajan, N., Liberman, Z. & Wynn, K. (2013). Not like me = bad: Infants prefer those who harm dissimilar others. *Psychological Science, 24*(4), 589–94. https://doi.org/10.1177/0956797612457785

12 Kristensen, P. & Bjerkedal, T. (2007). Explaining the relation between birth order and intelligence. *Science, 316*(5832), 1717. https://www.science.org/doi/10.1126/science.1141493

13 Sulloway, F. J. (2007). Birth order and intelligence. *Science, 316*(5832), 1711–12. https://doi.org/10.1126/science.1144749

14 Andersen, S., Ertaç, S., Gneezy, U., Hoffman, M., & List, J. A. (2011). Stakes matter in ultimatum games. *American Economic Review, 101*(7), 3427–39.
The researchers report that 20,000 rupees, the highest stake used, is equivalent to 1,600 hours' work at local rates. Assuming a moderately conservative $10 per hour (a little above the US national minimum wage of $7.25, but below that applied by many individual states) gives us the $16,000 figure.

15 Chiusano, A. (2023, July 17). The 25 biggest college football stadiums in the country. *NCAA*. https://www.ncaa.com/news/football/article/2018-07-30/25-biggest-college-football-stadiums-country

16 Sheffer, L., Loewen, P. J., Walgrave, S., Bailer, S., Breunig, C., Helfer, L., . . . & Vliegenthart, R. (2023). How do politicians bargain? Evidence from ultimatum games with legislators in five countries. *American Political Science Review, 117*(4), 1429–47. https://doi.org/10.1017/S0003055422001459

17 Levine, M., Prosser, A., Evans, D. & Reicher, S. (2005). Identity and emergency intervention: How social group membership and inclusiveness of group boundaries shape helping behavior. *Personality and Social Psychology Bulletin, 31*(4), 443–53. https://doi.org/10.1177/01461672042716

18 Harding, S. (2014). *The Street Casino.* Policy Press. p.64.
19 Ibid. p.191.
20 Ibid. p.258.
21 Ibid. p.64.
22 https://en.wikipedia.org/wiki/Peckham_Boys
23 Marsavelski, A., Sheremeti, F. & Braithwaite, J. (2018). Did nonviolent resistance fail in Kosovo? *The British Journal of Criminology, 58*(1), 218–36. https://doi.org/10.1093/bjc/azx002
24 https://www.youtube.com/watch?v=Braqs43CnmU
25 Marsavelski, A., Sheremeti, F. & Braithwaite, J. (2018). Did nonviolent resistance fail in Kosovo?
26 Ibid.
27 Clark, H. (2000). *Civil Resistance in Kosovo.* Pluto Press.
28 Ambridge, B., Doherty, L., Maitreyee, R., Tatsumi, T., Zicherman, S., Pedro, P. M., . . . & Fukumura, K. (2021). Testing a computational model of causative overgeneralizations: Child judgment and production data from English, Hebrew, Hindi, Japanese and K'iche'. *Open Research Europe, 1*(1). https://doi.org/10.12688/openreseurope.13008.2
29 McCauley, S. M., Bannard, C., Theakston, A., Davis, M., Cameron-Faulkner, T. & Ambridge, B. (2021). Multiword units lead to errors of commission in children's spontaneous production: 'What corpus data can tell us?' *Developmental Science, 24*(6), e13125. https://doi.org/10.1111/desc.13125
30 Ambridge, B., McCauley, S., Bannard, C., Davis, M., Cameron-Faulkner, T., Gummery, A. & Theakston, A. (2023). Uninversion error in English-speaking children's wh-questions: Blame it on the bigrams? *Language Development Research, 3*(1), 121–55. https://doi.org/10.34842/2023.641

7. Underdog

1 Public domain Library of Congress version at https://www.loc.gov/item/06016644/
2 Public domain version at https://penelope.uchicago.edu/Thayer/E/Roman/Texts/Strabo/17A3*.html#ref178
3 https://www.surlalunefairytales.com/oldsite/cinderella/marianroalfecox/etext.html

NOTES

4 Perrault, C. (1969). 'Cinderella'. *Perrault's Fairy Tales.* Trans. A. E. Johnson. Dover Publications.
5 https://en.wikipedia.org/wiki/The_Ugly_Duckling
6 Rice, S. (2016, May 17). Leicester will receive less Premier League prize money than Arsenal, Tottenham, Man United and Man City. *Independent.* https://www.independent.co.uk/sport/football/premier-league/leicester-to-receive-less-prize-money-than-arsenal-tottenham-manchester-united-and-man-city-a7031276.html
7 Hendricks, S. (2016, May 4). 5,000 to 1! Just how slim was Leicester City's slim chance of winning the English Premier League? *Slate.* https://slate.com/culture/2016/05/leicester-city-was-a-5000-to-1-underdog-how-big-of-an-underdog-is-that.html
8 Christoph (2023, June 14). The miracle of Leicester City in 2015/16 – Retro analysis. *Football Explained.* https://footballxplained.de/leicester-city-analysis-2015-16/
9 Burke, M. J., Romanella, S. M., Mencarelli, L., Greben, R., Fox, M. D., Kaptchuk, T. J., . . . & Santarnecchi, E. (2022). Placebo effects and neuromodulation for depression: a meta-analysis and evaluation of shared mechanisms. *Molecular Psychiatry, 27*(3), 1658–66. https://doi.org/10.1038/s41380-021-01397-3
10 Hurst, P., Schipof-Godart, L., Szabo, A., Raglin, J., Hettinga, F., Roelands, B., . . . & Beedie, C. (2020). The placebo and nocebo effect on sports performance: a systematic review. *European Journal of Sport Science, 20*(3), 279–92. https://doi.org/10.1080/17461391.2019.1655098
11 Gentrup, S., Lorenz, G., Kristen, C. & Kogan, I. (2020). Self-fulfilling prophecies in the classroom: Teacher expectations, teacher feedback and student achievement. *Learning and Instruction, 66,* 101296. https://doi.org/10.1016/j.learninstruc.2019.101296
12 Kahneman, D., Knetsch, J. L. & Thaler, R. H. (1991). Anomalies: The endowment Effect, loss aversion, and status quo bias. *Journal of Economic Perspectives, 5*(1): 193–206. doi:10.1257/jep.5.1.193
13 https://namelix.com
14 https://www.theo2.co.uk/assets/doc/FB-main-menu.pdf
15 Green, J. (2013, October 28). Jobs house added as 'historic resource'. *The Mercury News.* https://www.mercurynews.com/2013/10/28/jobs-house-added-as-historic-resource/

16 https://www.youtube.com/watch?v=XrhmepZlCWY
17 Waugh, R. (2014, December 9). Apple's Steve Wozniak debunks one of the biggest myths about firm's early days. *Metro*. https://metro.co.uk/2014/12/09/apples-steve-wozniak-debunks-one-of-the-biggest-myths-about-firms-early-days-4980457/
18 Siltanen, R. (2011, December 14). The real story behind Apple's 'Think Different' campaign. *Forbes*. https://www.forbes.com/sites/onmarketing/2011/12/14/the-real-story-behind-apples-think-different-campaign/?sh=67e3078a62ab
19 Cook, D. (2023, March 24). Better growth stock: Apple vs Microsoft. *The Motley Fool*. https://www.fool.com/investing/2023/03/24/better-growth-stock-apple-vs-microsoft/
20 United States Senate (1955). *Hearings before the committee on banking and currency, volume 3, part 1*. United States Government Printing Office (available online via Google Books).
21 Barsow, D., Craig, S. & Buettner, R. (2018, October 2). Trump engaged in suspect tax schemes as he reaped riches from his father. *The New York Times*. https://www.nytimes.com/interactive/2018/10/02/us/politics/donald-trump-tax-schemes-fred-trump.html
22 Flitter, E. (2016, July 17). Art of the spin: Trump bankers question his portrayal of financial comeback. *Reuters*. https://www.reuters.com/article/idUSKCN0ZX0GO/
23 McDonald, J., Karol, D. & Mason, L. (2019, January 17). Many voters think Trump's a self-made man. What happens when you tell them otherwise? *Politico*. https://www.politico.com/magazine/story/2019/01/17/many-voters-think-trumps-a-self-made-man-what-happens-when-you-tell-them-otherwise-224019
24 Michael, C. & agencies (2023, December 6). Trump says he will be a dictator only on 'day one' if elected president. *The Guardian*. https://www.theguardian.com/us-news/2023/dec/06/donald-trump-sean-hannity-dictator-day-one-response-iowa-town-hall
25 This Underdog story has a subplot. Jyoti's single was released on the Chrysalis label. Chrysalis was founded by Chris Wright, who attended the same school as my dad, where he was – at least as my dad tells the story – mocked as 'Goof Wright'. Wright got his big break when he funded the recording of Jethro Tull's debut album with money he didn't have, but went on to sell his label for more than $100 million:

NOTES

Barnes, M. (2023, December 18). 'We couldn't get a deal for Jethro Tull. The one person interested would only sign them if they dropped the flute player': How Chrysalis rose from a booking agency to a leading prog record label. *Louder Sound.* https://www.loudersound.com/features/chris-wright-chrysalis

26 Lerner, M. (1980). *The Belief in a Just World: A Fundamental Delusion.* New York: Plenum.

27 Hafer, C. (2000). Do innocent victims threaten the belief in a just world? Evidence from a modified stroop task. *Journal of Personality and Social Psychology,* 79(2), 165–73. https://doi.org/10.1037/0022-3514.79.2.165

28 Denke, C., Rotte, M., Heinze, H. J. & Schaefer, M. (2014). Belief in a just world is associated with activity in insula and somatosensory cortices as a response to the perception of norm violations. *Social Neuroscience,* 9(5), 514–21. https://doi.org/10.1080/17470919.2014.922493

8. Sacrifice

1 Genesis 22:2–12, New International Version.
2 https://quoteinvestigator.com/2016/05/05/brothers/
3 Tomasello, M. (2020). The moral psychology of obligation. *Behavioral and Brain Sciences,* 43, e56. https://doi.org/10.1017/S0140525X19001742
4 Ibid.
5 Whitehouse, H. (2018). Dying for the group: Towards a general theory of extreme self-sacrifice. *Behavioral and Brain Sciences,* 41, e192. https://doi.org/10.1017/S0140525X18000249
6 Most of the details and quotations from the experiment are taken from a transcription of a webinar given by Dr Tonya Foreman, a professor of psychiatry at the University of North Carolina Chapel Hill:
https://nceedus.org/wp-content/uploads/2023/07/Lessons-from-History_-Minnesota-Starvation-Experiment-Webinar-Transcript_Final.pdf
Some are from:
Guetzkow, H. S. & Bowman, P. H. (1946). Men and hunger: A psychological manual for relief workers. Brethren Publishing House.

which is available from the Open Library: https://openlibrary.org/books/OL6499821M/Men_and_hunger
7 Ibid.
8 This section is based on one of my columns for *The Observer*: Ambridge, B. (2017, May 14). Are you as environmentally friendly as you think? Personality quiz. *The Observer*. https://www.theguardian.com/lifeandstyle/2017/may/14/are-you-as-environmentally-friendly-as-you-think-personality-quiz
The original study on which the column was based is: Binder, M. & Blankenberg, A. K. (2017). Green lifestyles and subjective well-being: More about self-image than actual behavior? *Journal of Economic Behavior & Organization, 137*, 304–23. https://doi.org/10.1016/j.jebo.2017.03.009
9 Solnit, R.,(2021, August 23). Big oil coined 'carbon footprints' to blame us for their greed. Keep them on the hook. *The Guardian*. https://www.theguardian.com/commentisfree/2021/aug/23/big-oil-coined-carbon-footprints-to-blame-us-for-their-greed-keep-them-on-the-hook
10 Fibieger Byskov, M. (2019, January 10). Climate change: focusing on how individuals can help is very convenient for corporations. *The Conversation*. https://theconversation.com/climate-change-focusing-on-how-individuals-can-help-is-very-convenient-for-corporations-108546
11 Silke, A. (2015). Understanding suicide terrorism: Insights from psychology, lessons from history. In J. Pearse (ed.), *Investigating Terrorism*, (pp.169–79). Wiley.
12 Purvis, J. (2020, June 4). Emily Davison: the suffragette martyr. *History Extra*. https://www.historyextra.com/period/20th-century/emily-davison-the-suffragette-martyr/

9. Hole

1 This recipe is based on Blake Snyder's Dude with a Problem (https://savethecat.com/dude-with-a-problem) and Kurt Vonnegut's Man in a Hole:
Jones, J. (2014, February 18). Kurt Vonnegut diagrams the shape of all stories in a master's thesis rejected by U. Chicago. *Open Culture*. https://www.openculture.com/2014/02/kurt-vonnegut-masters-thesis-rejected-by-u-chicago.html

NOTES

2   In fact, there are so many fantastic real-life man-in-a-hole stories, I couldn't fit them all in! Here's a particularly amazing one about a man who singlehandedly saved the world from nuclear Armageddon: https://twitter.com/drg1985/status/1585524746799796224

3   Levin, S. (2018, July 16). Elon Musk calls British diver in Thai cave rescue 'pedo' in baseless attack. *The Guardian.* https://www.theguardian.com/technology/2018/jul/15/elon-musk-british-diver-thai-cave-rescue-pedo-twitter

4   This quotation is taken from an interview in the film adaptation of Joe Simpson's book, *Touching the Void* (with the same title).

5   Rees, J. (2019, November 16). 'By the end I'd lost me': Joe Simpson, mountaineer and writer – interview. *The Arts Desk.* https://theartsdesk.com/theatre/end-i'd-lost-me-joe-simpson-mountaineer-and-writer-interview

6   Valentish, J. (2022, January 16). Touching the Void: climber Joe Simpson on the 'feelgood' show inspired by his survival. *The Guardian.* https://www.theguardian.com/stage/2022/jan/17/touching-the-void-climber-joe-simpson-on-the-feelgood-show-inspired-by-his-survival

7   Guise, T. (2020, May 22). Joe Simpson recounts one of mountaineering's greatest survival stories. *The Red Bulletin.* https://www.redbull.com/gb-en/theredbulletin/joe-simpson-touching-the-void-interview

8   Del Vecchio, M., Kharlamov, A., Parry, G. & Pogrebna, G. (2018). The Data science of Hollywood: Using emotional arcs of movies to drive business model innovation in entertainment industries. *arXiv preprint arXiv:1807.02221.* https://arxiv.org/abs/1807.02221

9   Reagan, A. J., Mitchell, L., Kiley, D., Danforth, C. M. & Dodds, P. S. (2016). The emotional arcs of stories are dominated by six basic shapes. *EPJ data science, 5*(1), 1–12. https://doi.org/10.1140/epjds/s13688-016-0093-1

10 Di Pellegrino, G., Fadiga, L., Fogassi, L., Gallese, V. & Rizzolatti, G. (1992). Understanding motor events: a neurophysiological study. *Experimental Brain Research, 91,* 176–80. https://doi.org/10.1007/BF00230027

11 Filimon, F., Nelson, J. D., Hagler, D. J. & Sereno, M. I. (2007).

Human cortical representations for reaching: mirror neurons for execution, observation, and imagery. *Neuroimage, 37*(4), 1315–28. https://doi.org/10.1016/j.neuroimage.2007.06.008

12 Ramachandran, V. (2023). Mirror neurons and imitation learning as the driving force behind the great leap forward in human evolution. *Edge*. https://www.edge.org/conversation/vilayanur_ramachandran-mirror-neurons-and-imitation-learning-as-the-driving-force

13 Vittorio, G. and Guerra, M. (2020). *The Empathic screen: Cinema and neuroscience*. Oxford University Press.

14 Turvey, M. (2020). Mirror Neurons and Film Studies: A Cautionary Tale from a Serious Pessimist. *Projections, 14*(3), 21–46. http://doi.org/10.3167/proj.2020.140303

15 Engert, V., Linz, R., & Grant, J. A. (2019). Embodied stress: The physiological resonance of psychosocial stress. *Psychoneuroendocrinology, 105*, 138–46. https://doi.org/10.1016/j.psyneuen.2018.12.221

16 Verona, E., Patrick, C. J., Curtin, J. J., Bradley, M. M. & Lang, P. J. (2004). Psychopathy and physiological response to emotionally evocative sounds. *Journal of Abnormal Psychology, 113*(1), 99. https://doi.org/10.1037/0021-843X.113.1.99

17 Mastroianni, A. M. & Gilbert, D. T. (2023). The illusion of moral decline. *Nature, 618*, 782–9. https://doi.org/10.1038/s41586-023-06137-x

18 Rasmi, A. (2019, October 10). Only 1% of Brits cared much about the EU before the 2016 Brexit vote. *Quartz*. https://qz.com/1725402/only-5-percent-of-brits-cared-about-the-eu-before-brexit

19 Kessler, G. (2019, March 22). The Iraq War and WMDs: An intelligence failure or White House spin? *The Washington Post*. https://www.washingtonpost.com/politics/2019/03/22/iraq-war-wmds-an-intelligence-failure-or-white-house-spin/

20 Martinez, G. & Abrams, A. (2019, February 15). Trump repeated many of his old claims about the border to justify the state of emergency. Here are the facts. *Time*. https://time.com/5530506/donald-trump-emergency-border-fact-check/

21 Schmidt, J. (2022, September 5). California forests hit hard by

NOTES

wildfires in the last decade. *Wildfire Today.* https://wildfiretoday.com/2022/09/05/california-forests-hit-hard-by-wildfires-in-the-last-decade/#:~:text=In%20the%20following%20decade%20(2012,7.9%20million%20acres%20(24.7%25

22 Boehm, S. (2022). *State of Climate Action 2022.* World Resources Institute. https://doi.org/10.46830/wrirpt.22.00028

23 McKie, R. (2022, October 30). Cop27 climate summit: window for avoiding catastrophe is closing fast. *The Guardian.* https://www.theguardian.com/environment/2022/oct/30/cop27-climate-summit-window-for-avoiding-catastrophe-is-closing-fast

24 Horton, H. (2022, October 26). Atmospheric levels of all three greenhouse gases hit record high. *The Guardian.* https://www.theguardian.com/environment/2022/oct/26/atmospheric-levels-greenhouse-gases-record-high

25 Jolly, J. (2022, October 27). Carbon emissions from energy to peak in 2025 in 'historic turning point', says IE. *The Guardian.* https://www.theguardian.com/environment/2022/oct/27/carbon-emissions-to-peak-in-2025-in-historic-turning-point-says-iea

26 Griffith, S. (2022). *Electrify: An optimist's playbook for our clean energy future.* MIT Press.

27 Cormier, Z. (undated). Turning carbon emissions into plastic. *BBC Earth.* https://www.bbcearth.com/news/turning-carbon-emissions-into-plastic

28 Simon, M. (2021, October 20). Microplastics may be cooling – and heating – Earth's climate. *Wired.* https://www.wired.com/story/microplastics-may-be-cooling-and-heating-earths-climate/

29 https://www.drawdown.org/solutions/table-of-solutions

30 Guillot, L. (2020, September 16). How recycling is killing the planet. *Politico.* https://www.politico.eu/article/recycling-killing-the-planet/

31 Dasandi, N., Graham, H., Hudson, D. *et al.* (2022). Positive, global, and health or environment framing bolsters public support for climate policies. *Communications Earth and Environment, 3,* 239 (2022). https://doi.org/10.1038/s43247-022-00571-x

10. The Stories of Your Life

1. Garde, D. & Saltzman, J. (2020, November 10). The story of mRNA: How a once-dismissed idea became a leading technology in the Covid vaccine race. *Stat*. https://www.statnews.com/2020/11/10/the-story-of-mrna-how-a-once-dismissed-idea-became-a-leading-technology-in-the-covid-vaccine-race/

ACKNOWLEDGEMENTS

Ause Abdelhaq at Pan Macmillan has been not only a fantastic editor, but a words-can't-explain-how-enthusiastic champion of the whole project: 'It's as though you went into my brain and plucked out everything I've been thinking about for at least three to four years', he told me after reading the proposal. Before Ause joined, Matt Cole bought the book for Pan Macmillan (beating off strong competition, I should add!) and was my editor for a few months before crossing the table to become an agent (with Mike Harpley holding the editorial fort before Ause took over). Rebecca Needes was the senior desk editor for the book, with Fraser Crichton copy editing. My agent, Sally Hollway at Felicity Bryan Associates, did a heroic job of refining and selling the proposal, persevering through a process that (through no fault of Sally's, I should make clear) took several years. Juliet Garcia – then also at Felicity Bryan – deserves special thanks for almost singlehandedly rescuing the proposal when it looked in serious danger of going under.

I will be forever grateful to Daniel Crewe and Nick Sheerin, both then at Profile Books, who kickstarted my writing career, along with Profile's founder, Andrew Franklin. This isn't an academic book, but I would never have been in a position to write it without the support of my academic mentors: first my PhD Supervisors, Mike Tomasello, Elena Lieven and Anna

THE STORIES OF YOUR LIFE

Theakston at the University of Manchester, then Julian Pine and Caroline Rowland at the University of Liverpool.

Thanks are due to all the people who took time out of their busy schedules to be interviewed for this book: Jon Cole, Laurence Alison, Karl Bushby, Jyoti Mishra, Peter Moore, Simon Harding and Tom Hartley. I hope I haven't misquoted or misrepresented any of you. If I have, you know where to find me. A book like this one relies on a huge number of academic and non-academic sources, and I thank everyone whose work made it possible. Amongst all of these sources, one deserves special mention: Christopher Booker's *The Seven Basic Plots: Why We Tell Stories*, which was the original source for many of the masterplot recipes set out here.

I dedicate this book to Louise and our children – there's no way to say this without it sounding cheesy, but it's thanks to the three of you that the stories of my life have been overwhelmingly happy ones.

Ben Ambridge
Sale, April 2024